KT-171-283

WINCHESTER SCHOOL OF MISSION

08144

THE NATURAL HISTORY OF SELBORNE

GILBERT WHITE (1720–93) was born in Selborne and attended school in Basingstoke. Even when he was young he made notes on natural history. At the age of nineteen he went to Oriel College, Oxford, where he became a Fellow in 1743. He spent much of the next fifteen years travelling around England, but always returned to his beloved Selborne. In 1751 he began keeping his 'Garden Kalendar' and later, his 'Naturalist's Journal'. *The Natural History of Selborne* is based on his correspondence with two distinguished naturalists, Thomas Pennant and Daines Barrington. The book became extremely popular, and is reputedly the fourth most published book in the English language.

RICHARD MABEY is well known as a writer on countryside matters. His best known books are *The Frampton Flora*, *Food for Free*, *In a Green Shade*, *The Common Ground*, *The Unofficial Countryside* and *The Flowering of Britain* (with Tony Evans). These last two were made into BBC documentaries. His biography of Gilbert White won the 1986 Whitbread Biography of the Year Award. He also writes regularly for publications such as the *Independent*, the *Sunday Times*, *Modern Painters*, *Country Living* and *Nature*, and is a columnist for *BBC Wildlife*.

Resource Room
Church House
9 The Close
WINCHESTER
SO23 9LS
Tel: 01962 624760

PENGUIN CLASSICS

GILBERT WHITE

THE NATURAL HISTORY
OF SELBORNE

EDITED WITH AN

INTRODUCTION AND NOTES BY

RICHARD MABEY

PENGUIN BOOKS

PENGUIN BOOKS

Published by the Penguin Group
Penguin Books Ltd, 27 Wrights Lane, London W8 5TZ, England
Penguin Putnam Inc., 375 Hudson Street, New York, New York 10014, USA
Penguin Books Australia Ltd, Ringwood, Victoria, Australia
Penguin Books Canada Ltd, 10 Alcorn Avenue, Toronto, Ontario, Canada M4V 3B2
Penguin Books (NZ) Ltd, 182–190 Wairau Road, Auckland 10, New Zealand

Penguin Books Ltd, Registered Offices: Harmondsworth, Middlesex, England

First published 1788–9
Published in the Penguin English Library 1977
Reprinted in Penguin Classics 1987
7 9 10 8 6

Introduction and notes copyright © Richard Mabey, 1977
All rights reserved

The maps which precede the text are taken from Anthony
Rye, *Gilbert White and his Selborne* (William Kimber,
1970), and are reproduced by kind permission of the
publishers.

Printed in England by Clays Ltd, St Ives plc
Set in Monotype Bembo

Except in the United States of America, this book is sold subject
to the condition that it shall not, by way of trade or otherwise, be lent,
re-sold, hired out, or otherwise circulated without the publisher's
prior consent in any form of binding or cover other than that in
which it is published and without a similar condition including this
condition being imposed on the subsequent purchaser

CONTENTS

INTRODUCTION

On 24 November 1822, on one of his rides through the back-waters of Hampshire, William Cobbett passed a turning for a village whose name touched on a nerve:

I forgot to mention, that, in going from Hawkley to Greatham, the man, who went to show me the way, told me at a certain fork, 'that road goes to *Selborne*.' This put me in mind of a book, which was once recommended to me, but which I never saw, entitled '*The History and Antiquities of Selborne*,' (or something of that sort) written, I think, by a parson of the name of *White*, brother of Mr *White*, so long a bookseller in Fleet-street. This parson had, I think, the living of the parish of Selborne. The book was mentioned to me as a work of great curiosity and interest. But, at that time, the THING was biting *so very sharply* that one had no attention to bestow on antiquarian researches. Wheat at 39s. a quarter, and South-Down ewes at 12s. 6d. have so weakened the THING's jaws and so filed down its teeth, that I shall now certainly read this book if I can get it. [30]*

Cobbett did not take the fork on that occasion. But one of his readers sent him a copy of *The Natural History and Antiquities of Selborne, in the County of Southampton*, by the Reverend Gilbert White, M.A., which had been published thirty-four years before. True to his word, Cobbett read it, and was sufficiently affected to make a special trip to White's parish the following autumn (1823). He loved Selborne, which was 'precisely what is described by Mr White . . . Nothing can surpass in beauty these dells and hillocks and hangers.' And like every pilgrim since, he searched out the 'curiosities' which White had written about with such easy vividness, the zig-zag path,

---

*Figures in square brackets refer to entries in the Bibliography. Quotations that are *not* annotated in this way come from the text of the *Natural History*.

the hollow ways, and the churchyard with its ancient yew. 'The land,' he wrote, 'is good, all about it.'

Nothing captures more perfectly the anomaly of White's book than the transition between the bluff superiority of Cobbett's first note, and the respectful passion of his second. Both, in their own ways, are entirely reasonable responses. Considered at its face value, the book *was* something of a phenomenon. How could the jottings of a rural vicar (or was he just a curate?) about the goings-on of the non-human inhabitants of an unknown Hampshire village have such a mesmeric effect on all who read them? How was it that a collection of letters on local natural history – for that is all there is to the book – written with wilful parochialism at a time of social revolution in France and agricultural upheaval at home, became the fourth most published book in the English language? There had already been at least four editions of the book by the time Cobbett was writing. By the 1970s the new editions and translations numbered in excess of two hundred. The early cheap versions were carried all over the world by expatriate Englishmen, sick for the dells and hillocks of home. It is worshipped by American naturalists, who have never seen half the creatures White talks about, and by the Japanese, who have seen even fewer.

It's no wonder that the search for a key to White's unique appeal has preoccupied his critics and editors – sometimes to the point of obsession, as if there were some stylistic mannerism, some trick of words, which, properly decoded, would reveal the secret. There is not, of course; and *Selborne's* strength lies more in its refreshing ingenuousness than in any contrivance. But though it won't, thankfully, be explained away, it must be seen in perspective. More than any other single book it has shaped our everyday view of the relations between man and nature. I say 'everyday' without in any way meaning to belittle White's scientific contributions. These were considerable, as we shall see, particularly in the area of observational method.

But their impact on scientific theory was small, and they were soon overshadowed by the discoveries of giants like Darwin and Mendel.

White's contribution was more personal, in both senses. He was perhaps the first writer to talk of animals – and particularly birds – as if they conceivably inhabited the same universe as human beings. In White's chosen patch, the village of Selborne, they did so literally. Listen to this passage on the swift. White is writing in the summer of 1774 from his house in the High Street, just across the village square from the church where he was curate:

It is a most alert bird, rising very early, and retiring to roost very late; and is on the wing in the height of summer at least sixteen hours. In the longest days it does not withdraw to rest till a quarter before nine in the evening, being the latest of all day birds. Just before they retire whole groups of them assemble high in the air, and squeak, and shoot about with wonderful rapidity. But this bird is never so much alive as in sultry thundry weather, when it expresses great alacrity, and calls forth all its powers. In hot mornings several, getting together in little parties, dash round the steeples and churches, squeaking as they go in a very clamorous manner; these, by nice observers, are supposed to be males, serenading their sitting hens; and not without reason, since they seldom squeak till they come close to the walls or eaves, and since those within utter at the same time a little inward note of complacency.

(*Letter XXI to Daines Barrington*)

This is a new note in nature writing – and not just because of the accuracy and percipience of the observations. White is plainly talking from his own experience (he had even built one of those steeples himself). His swifts are real, not puppets – 'bundles of responses' – or links in some taxonomic chain. They are living birds in a living and closely observed situation. With no more than a hint of anthropomorphism, White suggests that their lives have a richness and rhythm of their own. And by describing them on a summer evening in a particular

English village, he is able to awake in us the possibility of a human response to them.

If this seems unexceptional today, it is partly a result of White's own pioneering. It is important to remember that he was living at the tail-end of the Enlightenment, when science was heady with its own progress and dominated by systems-builders like Ray and Linnaeus. It was also a time marked by an enormous influx of exotic plants and animals, sent back to Britain by explorers. As a consequence, most naturalists hid away in their laboratories, trying to fit these oddities (which were mostly distorted corpses when they arrived) into some coherent scheme of things. It was a typical contemporary response, a desire to drill some order into the disarrayed ranks of Creation.

White was more interested in the relationships between living animals than in the relationships between their taxonomically significant parts. (He wrote to Daines Barrington that 'system should be subservient to, not the main object of the pursuit.')

But this is no more than rephrasing the problem. How was this unexceptional country parson able to go against the whole tenor of science in his time, and rise to such a position of eminence in both natural history and literature? What moved him to keep such meticulous records of his observations, and craft them – with more purposefulness than is obvious at first sight – into a book of a kind which was quite new in the natural sciences? And what was the bond that drew him back so persistently to that tenacious Hampshire village?

Selborne – 'a stream through the sallows'. It is impossible to talk about White outside the context of the village in which he lived and died. Selborne was the source of his inspiration, and has become a memorial almost equal in stature to the book it inspired. When you look at the village – and it has changed very little in the last two hundred years – it is not difficult to see why White loved it, and how it was so easily and winningly

immortalized. In many ways it is a thoroughly ordinary settle-
ment: a single main street with a few side lanes, lined with
cottages that are neither dull nor excessively picturesque. There
is a green next to the church (the Plestor, which had its own
maypole in White's day), a couple of pubs, and an old butcher's
shop across the street from White's house. But Selborne lies
cupped in a bowl formed by a confluence of soils and streams,
and the countryside immediately around it is, without exaggeration, idyllic. A steep beech-clad hanger rises just behind the
village. Rich pastures spread out towards the north-east,
and beyond the church there is a marvellous jumble of dells and
copses ringed by a perpetual clamour of rooks. In half an hour's
stroll you can pass through oak forest, chalk downland, sandy
heaths, water meadows, hop fields, and a pattern of agriculture
that seems scarcely to have been touched by the revolutions of
the last two centuries. Selborne, in short, is the landscape of the
pastoral dream made flesh: cosseted, balanced, endurable, a
condensation into a single parish of almost every type and mood
of scenery to be found in lowland England.

That this is the place where White lived is, in one sense, the
only important fact in his biography. He was born in the
vicarage on the 18 July 1720, and died a hundred yards away in
'The Wakes' on 26 June 1793. He never married, and remained
a country clergyman all his life. Everything that happened to
him, by decision or default, seemed designed to preserve his
single-minded commitment to his native village. His father,
John, was a barrister, but he was an exception in a family otherwise almost exclusively dedicated to the Church. Gilbert's
grandfather (also called Gilbert) was vicar of Selborne, his
mother a clergyman's daughter, and most of his uncles and
nephews had livings in neighbouring parishes. The pressure on
Gilbert to follow in their footsteps must have been considerable,
though at no stage did he show any dissatisfaction with his
calling.

In 1727, Gilbert's grandfather died, and the whole family

moved across the Plestor to the house now known as 'The Wakes'. Gilbert went to schools in Farnham and Basingstoke, and even at this comparatively early age began to keep spasmodic records of his natural history observations. Before he was eleven he was planting out his own trees in the grounds of The Wakes.

At the age of nineteen he went to Oriel College, Oxford. He seems to have been a thoroughly normal, perhaps even hearty, undergraduate, and whatever sensitivity he was to develop later in his life was certainly not apparent then. He rode, hunted, shot and fished with almost ferocious vigour. Ironically – given what was to become his consuming passion – he used to gun down small summer migrants to keep his hand in for the winter shoots.

He became a fellow of Oriel in 1743, and spent the next fifteen years travelling around England, returning continually, and with apparently increasing devotion, to his native village. He could never become vicar of Selborne (the living was with Magdalen College, not Oriel) and during this period he took up five temporary curacies, three of them at Selborne. His Oxford colleagues were forever tempting him to apply for positions at Oriel, but only once, apparently, did he succumb, when he unsuccessfully applied for the Provostship. His beloved village was too strong a magnet, and in 1761 he settled for the curacy of Farringdon. This village is only two miles from Selborne, and White was able to carry out his duties from The Wakes. He kept this position for twenty-four years, and in 1784, on the death of the incumbent at Selborne, finally became curate of his birthplace for the fourth time, a position which he held until his death.

He led, by all accounts, a quiet and relatively leisurely life. He never scrimped his clerical duties, but with only a few dozen marriages and burials to attend to a year, had plenty of free time to pursue his natural history. Nor did he ever permit his religious persuasions to intrude upon his reasoning. There is no trace in his writing of moralizing about 'brute creation', or of the mawkish search for natural parables that fill the books of the Victorian parson-naturalists. And as men of God go, he enjoyed his share of earthly pleasures; there is even a touch of the mannered aesthete about him. He was especially partial to composed views and Gothic follies. There is a delightful description in Letter XXXVIII to Daines Barrington of one of his friends discovering the echo in Galley Lane: 'At first he was much surprised, and could not be persuaded but that he was mocked by some boy; but, repeating his trials in several languages, and finding his respondent a very adroit polyglot, he then discerned the deception.' It is slightly surprising that White did not seek to preserve the echo by keeping the 'object' of the echo (a hop-kiln) clear of foliage, for the superficial topography of much of the village is the result of his landscape gardening. He built the famous zig-zag up the Hanger, and at the top a hermitage in which to hold summer picnics. One of the uninvited guests one evening was a nightjar: 'You will credit me, I hope, when I tell you that, as my neighbours were assembled in an hermitage on the side of a steep hill where we drink tea, one of these churn-owls came and settled on the cross of that little straw edifice and began to chatter, and continued his note for many minutes: and we were all struck with wonder to find that the organs of that little animal, when put in motion, gave a sensible vibration to the whole building!' It is likely that he found this visitation just as bewitching as those of the flighty Battie sisters, and of the one person it has been suggested he had a romantic attachment to – Hester ('Hecky') Mulso (she nicknamed him 'Whitey-bus'), the sister of White's close friend and mentor, John Mulso. But the relationship came to nothing, and as Rashleigh Holt-White [21] put it, White 'had but one mistress – Selborne'.

It's no real wonder, then, that White's first serious writings were concerned with the man-made, and took the form of a *Garden Kalendar*. This was a diary of the principal events in the horticultural year, which he began in 1751. But by 1768 he had expanded it into a more general 'Naturalist's Journal', which covered records and observations of the natural world as well as the garden. He was encouraged in this by the explorer and naturalist Daines Barrington, who that year had sent him one of the special diary blanks he had had printed up by Gilbert's brother, Benjamin. Gilbert immediately began filling it with his daily observations on weather, flowers and the comings and goings of birds, and continued almost without a break until four days before his death. The manuscript of these journals in the British Museum gives a fascinating insight into the development of White's interests. The early entries are disciplined and terse, each note on barometric pressure and 'trees first in leaf' neatly confined to its own narrow column. But as White's skills in observation and imaginative reasoning develop (and perhaps the first plans for a book start to form), so the records and observations start to trespass outside their columns.

Gilbert met Daines Barrington in Benjamin's bookshop in May 1769, and began to correspond with him one month later. But he had already started the formal correspondence which was to form the *Natural History* two years earlier, with the eminent zoologist Thomas Pennant, whom he had also met in Benjamin's shop. It's difficult to appreciate now what an intellectual lifeline correspondence must have been to country-dwellers in the eighteenth century. All road transport was slow and hazardous (White was additionally cursed with acute coach-sickness), and though letters went by the same routes, at least they could be sent more frequently. White complains more than once of the lack of neighbours with which to discuss natural history, and the pleading for information in some of his letters to his brother John in Gibraltar reveals an unashamed and impatient yearning for the exchange of knowledge and ideas.

The first mention of the possibility of a book based on White's journals and letters is in his note to Barrington of 12 April 1770 (p. 122). Yet there are clues which suggest White may have been thinking of publication for some time, and that, with a flash of editorial flair and prescience, had seen that a series of letters, datelined Selborne, would most accurately convey the essence of that parochial ecology which was his private passion and public discipline. The tone of the 'book' letters, for instance, is notably more formal and literary than that of the chatty notes he wrote to his friends on other subjects. He even tops and tails the collection of real letters that comprise the *Natural History* with mocked-up letters in the style of the rest of the book. And in 1773 he writes to a friend: 'Pray tell me over again the story of the swallow building on the dead owl's wing, and on the conch, etc; I think I could make good use of it.' [28] The information was supplied and subsequently worked into Letter XVIII to Barrington. These, I would suggest, are not the moves of some innocent rustic scribe, but of a shrewd working writer.

Nevertheless, eighteen years were to elapse between the first mention of the project and the delivery of the manuscript – a delay brought about by what White's friends regarded as his incurable procrastination, and by his curious insistence on compiling a dull and derivative appendix on the *Antiquities of Selborne*. (And perhaps by some discreet editing of those early letters?)

So, in December 1788, Selborne had its *genius loci* laid bare in print. And the immemorial quality of that village found itself reflected in a book which not only endured itself, but, by a kind of symbiosis, helped the village to endure too. One of the *Natural History*'s most reassuring qualities is the sense it gives of an unbroken line of continuity in the life of the countryside. Two of the lime trees that White planted to hide the blood-stained forecourt of the butcher's shop are still standing. Swifts still arrive in the village in the first week in May, and fly round the steeples in much the same numbers as were counted by White. Helleborines bloom on the Hanger, and golden saxifrage in damp ditches by the lanes. The thousand-year-old yew continues to have its barely perceptible annual growth reverently charted. (Even Cobbett put a tape round it, and found that 8 inches had been added to its 23 feet girth since White's day.) From our embattled modern country-side it is curiously heartening to look back and see that White, too, had to wrestle with the problems of heath fires and vandals and venison poachers. Even as I write this, southern England is in a state of extreme drought, and the Forestry Commission is broadcasting emergency fire warnings for Hampshire. Two hundred years ago to the day, White wrote in his journal for 30 April 1776: 'Birds silent for want of showers . . . no rain since the beginning of March.' Life goes on. Yet if White had a strong sense of the invariables of natural life, he also had a sense of occasion, and, even more, of place. He was one of the first documentarists, and if he was living today he would probably be making natural history films.

The place for which he had the greatest feel was, of course, Selborne. There were not many others that he really knew. Whilst Cook and Banks were botanizing on the other side of the world, and his correspondent Pennant was leading an expedition to the Hebrides, White was watching bats over his lawn. He never travelled further than Derbyshire, and was so overcome by the size of the Sussex Downs that he referred to them as that 'vast range of mountains'. He was bound to, and by, Selborne. It was his manor, and he surveyed it on horseback like a benevolent squire, attentive to the lives of all its inhabitants. Other men, more in tune with the expansiveness of the times, might gaze briefly at migrant birds in tropical skies. But White had watched them in their nests, and even, on occasion, held them in his hands. He had 'gently insinuated' field crickets from their holes on a grassy bank just three hundred yards behind the church. He knew the exact length of the bounds of his parish.

I have used that word 'parish' many times already. It seems to me crucial to understanding the impact of White's writing.

He was a parish priest. He talks in the very first sentence of his book of his desire to write a 'parochial history'. (David Elliston Allen [27], in a memorable phrase, has described Selborne as 'the secret, private parish inside each one of us'.) 'Parish' is a very laden concept. It has to do not just with geography and ecclesiastical administration, but with history and a system of loyalties. For most of us, it is the indefinable territory to which we feel we belong, which we have the measure of. Its boundaries are more the limits of our intimate allegiances than lines on a map. These allegiances have always embraced wild life as well as human, and show themselves no more explicitly than in the universal delight at the annual return of the 'village' birds – the swifts and swallows and martins. I don't think it is a coincidence that these were White's favourites. But his originality lies in the way he wedded a celebration of these thoroughly irrational parochial loyalties to sharp and dispassionate observation. It is this blend which is the quintessence of what we know as 'natural history', and which makes it a different kind of knowledge from biology. It has survived even into these days of scientific rigour. Town and village floras are still published, and birds still harried across county boundaries so that they can be added to local tallies. (There is even a recently published list of lichens for a village which dates back to Saxon times, and whose parish boundaries must have been established before this class of plants had even been named.) It hardly needs saying that these artificial territories have little scientific significance. But, lovingly watched and logged as they are, they help provide a meeting place for wild life and human feeling, and we must thank White for that.

His writing style was parochial too. In the very best sense of the word he was a gossip. He delighted in telling stories about his own and others' observations, though never about himself. He writes rivetingly about a raven dashed to her death as the tree in which she was nesting, and which she refused to desert, was felled; about an extraordinary shower of gossamer on the

district; about the moose which he made a special visit to Richmond to see, but 'was greatly disappointed, when I arrived at
the spot, to find that it died ... on the morning before'. He
examined it meticulously, but the poor beast was already
putrefying and White had to cut short his studies: 'From the
fore-feet to the belly behind the shoulder it measured three feet
and eight inches: the length of the legs before and behind
consisted a great deal in the *tibia*, which was strangely long; but
in my haste to get out of the stench I forgot to measure that
joint exactly.'

This sentence is pure White: eager, precise, unaffectedly
witty, and scrupulously honest. Nowhere in his work does he
trust absolutely in anything other than his own observations.
When he refers to another's opinions or records he invariably
makes it clear he is quoting. He is not afraid to rebuke scientists
as distinguished as Linnaeus and Scopoli for their occasional
gullibility and lapses of discipline. Writing to Daines Barrington about the contentious behaviour of the woodcock, he says:
'Faunists, as you observe, are too apt to acquiesce in bare
descriptions, and a few synonyms: the reason is plain; because
all that may be done at home in a man's study, but the investigation of the life and conversation of animals, is a concern
of much more trouble and difficulty, and is not to be attained
but by the active and inquisitive, and by those that reside
much in the country.'

White, out on his rounds with a notebook always in his
pocket, was all of these things, and this is why he was able to
make so many original contributions to British natural history.
He distinguished between our three summer leaf-warblers. He
discovered the harvest mouse, and, in Britain at least, the noctule bat. A hundred years before Darwin, he realized the crucial
role of worms in the formation of soil. He was one of the very
first naturalists to understand the significance of territory and
song in birds. The full list of his discoveries and original
insights would cover many pages.

But White's unique contribution to natural history lies less in the catalogue of species and concepts that he added to the textbooks, than in the style of his perception: *how* he saw, and how he communicated what he saw. Let me quote a rather lengthy passage about house martins to illustrate this:

About the middle of May, if the weather be fine, the martin begins to think in earnest of providing a mansion for its family. The crust or shell of this nest seems to be formed of such dirt or loam as comes most readily to hand, and is tempered and wrought together with little bits of broken straws to render it tough and tenacious. As this bird often builds against a perpendicular wall without any projecting ledge under, it requires its utmost efforts to get the first foundation firmly fixed, so that it may safely carry the superstructure. On this occasion the bird not only clings with its claws, but partly supports itself by strongly inclining its tail against the wall, making that a fulcrum; and thus steadied it works and plasters the materials into the face of the brick or stone. But then, that this work may not, while it is soft and green, pull itself down by its own weight, the provident architect has prudence and forbearance enough not to advance her work too fast; but by building only in the morning, and by dedicating the rest of the day to food and amusement, gives it sufficient time to dry and harden. About half an inch seems to be a sufficient layer for a day. Thus careful workmen when they build mud-walls (informed at first perhaps by this little bird) raise but a moderate layer at a time, and then desist; lest the work should become top-heavy, and so be ruined by its own weight. By this method in about ten or twelve days is formed an hemispheric nest with a small aperture towards the top, strong, compact, and warm; and perfectly fitted for all the purposes for which it was intended.

(Letter XVI to Daines Barrington)

This is a remarkable passage, not just because of White's uncanny ability to interpret the reasons behind birds' behaviour, but because of its *precision*. The exactness of his observations and of his language are inseparable. The martins, for example, do not just use 'grass', but 'little bits of broken straws', an image that is at once more sympathetically evocative of the energetic activities of these little 'martlets', and also more helpful in understanding how they actually put their nests together.

White has often been criticized for the lack of attention he pays in the book to his human parishioners. Richard Jefferies wrote that 'He knew the farmers and squires; he had access everywhere, and he had the quickest of eyes. It must ever be regretted that he did not leave a natural history of the people of his day ... The gallows-trees grew far too plentifully at the cross roads in those days, and the laws were inhuman, men were put to death like wild beasts: in fact they seemed to look on a man as a species of wolf that could only be tamed by stretching its neck.' [13] This scarcely sounds like the same world that White describes. Yet, to tell the truth, there is a kind of animal quality about the few humans that do cross his pages. They are curiosities, like the bat and the moose: a leper, a mad bee-boy, gypsies, a toad-witch. They are described objectively, but with little of the sympathy or respect that White shows for his favourite wild creatures. They are not met on their own terms.

There is no suggestion of malice here, or even of aloofness. By all accounts White was a sociable and easy-going man, and much liked by his parishioners. I think it is simply that, although he brought the natural world closer to the human, he did not take the next step, and see that ordinary human affairs were in their turn part of the natural scheme of things. And perhaps we should be glad that he didn't, for I doubt if the book could have possessed its enduringly peaceful quality if White had known what we now know about the unity and fragility of life on earth. For him the natural world was infinitely renewable and infinitely beneficent. Who else, on the day of the storming of the Bastille, would be writing a whole page in his diary on the nesting habits of the nightjar?

White had rescued natural history from the dusty cabinets of the taxonomists, given it a human scale and a human setting, *humanized* it. Yet what a double-edged word that is. It also meant that *Homo sapiens'* 'dominion over nature' was now social as well as intellectual and physical. Knowledge was still the supreme goal, and White had few scruples about sacrificing the lives that delighted him in order to gain more of it. Yet I believe he was beginning to be troubled by the kind of sympathies – call them sentiments if you wish – that did not come fully into the open until half a century later, in the writings of men like the poet John Clare. White loved his creatures, yet was unsure how to convey this emotion except by affectionately accurate description. He experiences regret, wonder, joy, grief, but is diffident about working these feelings into the details of what he observed. The force of this suppressed emotion helps, I think, to explain White's obsession with the problem of whether swallows and their kind returned *en masse* to Africa, or whether some of them spent the winter with us in 'hybernaculums'. He never resolved the problem, for he never found the evidence his high scientific principles demanded for the resolution of a question. But for thirty years he was preoccupied by migration and hibernation. He demanded records from his correspondents, poked about in likely holes in rocky lanes, and hired men to scour the winter bushes for sleeping birds. He raises the issue in more than half of the true letters to Pennant and Barrington, and in one exchange, it's clear that he is being partly driven by sheer frustration at his inability to fill this yawning gap in science:

When I used to rise in a morning last autumn and see the swallows and martins clustering on the chimnies and thatch of the neighbouring cottages, I could not help being touched with a secret delight, mixed with some degree of mortification: with delight to observe with how much ardour and punctuality those poor little birds obeyed the strong impulse towards migration, or hiding, imprinted on their minds by their great Creator; and with some degree of mortification, when I reflected that, after all our pains and inquiries, we are yet not quite certain to what regions they do migrate; and are still farther embarrassed to find that some do not actually migrate at all.

(*Letter XXIII to Thomas Pennant*)

But it is the admission of that 'secret delight' in the possibility of 'hiding' that lingers in my mind. And as the subject is aired in letter after letter, it begins to look as if White simply did not *want* the swifts and swallows and martins to go. Such was his loyalty to his village, and to these 'parish' birds that magnanimously shared its streets and steeples for the summer, that I think that in an unscientific corner of his heart he hoped that a few might stay there with him for the winter. And is it being too fanciful to read in his reference to Goree (an insignificant island off the West Coast of Africa; page 38) as one of the wintering quarters of his martlets, an example chosen because of the slight distastefulness of its name?

White never openly admitted feelings of this kind in his book, and the sadness is that two centuries of less gifted nature writers, looking to him as a model, have made concrete that apparent separation between fact and feeling. But it's doubtful if White himself – for all his restraint and self-effacement – saw objectivity and delight as contradictory. There is an entry on the arrival of the first swallow of summer in his Journal for 1768, whose excitement no cold Latin objectification can disguise. It reads simply:

'April 13: Hirundo domestica!!!'

BIBLIOGRAPHY

WHITE'S WORKS

THERE has never been a complete edition of White's writings, nor of the journals which were the basic resource from which he built up his papers and letters. The latter comprise over nine thousand daily entries between the years 1768 and 1793. There is one short selection in print.

[1] JOHNSON, WALTER (ed.), *Gilbert White's Journals*, 1931; reprinted Routledge & Kegan Paul, 1970.

Two years after White's death in 1793, Dr John Aikin (1747–1822) went through the manuscript of the journals and put together a record of the first occurrences of natural phenomena through the year. Side by side with White's dates are placed those of William Markwick (see [6]), who lived in Sussex. Also included in these gleanings were various more substantial passages from the journals, which Aikin classed together as 'Observations on Various Parts of Nature'.

[2] WHITE, GILBERT, *A Naturalists' Calendar, and Observations on Various Parts of Nature*, edited by Dr John Aikin, 1795.
[3] JESSE, EDWARD (ed.), *Gleanings in Natural History*, 1834. Contains substantially the same text.

There is also the *Garden Kalendar*, which was the diary White kept before his 'Naturalist's Journal'. Some of this was included in Bell [12] and the bulk in Sharpe [15]. The only correct and complete edition is a modern facsimile:

[4] WHITE, GILBERT, *The Garden Kalendar*, edited by John Clegg, The Scolar Press, 1976.

In addition to the letters to Thomas Pennant and Daines Barrington, which comprise the *Natural History*, there are the *Antiquities of Selborne* (rarely included in modern editions), some poems (first printed in an edition dated 1813), ten letters to Robert Marsham, in Norfolk, which were published in the 'Transactions of the Norfolk and Norwich Naturalists' Society' in autumn 1875, and first reprinted in the second edition of Harting [11].

BIBLIOGRAPHY

EDITIONS OF *THE NATURAL HISTORY OF SELBORNE*

Many of the most distinguished naturalists of their day have edited the *Natural History*, and their notes and introductions provide a record not only of changing attitudes in the natural sciences, but of changing fashions in nature writing. The list below is of some of the more important editions, arranged in chronological order:

[5] WHITE, GILBERT, *The Natural History and Antiquities of Selborne, in the County of Southampton: with Engravings, and an Appendix*, published by Benjamin White & Son, London, December 1788 (though the date of publication on the title-page is 1789). The first edition. A facsimile was published by the Scolar Press, London, in 1976.

[6] MARKWICK, WILLIAM (ed.), 1802. Contains the *Natural History*, plus the *Naturalist's Calendar* and the *Observations* (see [2] above).

[7] JARDINE, Sir WILLIAM (ed.), 1829. An improved version of this edition, containing the *Antiquities* and a map of the district, was published in 1853.

[8] DOVER, Lady GEORGIANA (ed.), 1833. Specially 'arranged for young persons', and later published under the auspices of the Society for the Promotion of Christian Knowledge.

[9] BENNET, EDWARD TURNER (ed.), 1837. A very complete edition, containing notes by Bell, Daniell, Owen and Yarrell.

[10] JESSE, EDWARD (ed.), 1851. With additional notes by Jardine.

[11] HARTING, JAMES EDMUND (ed.), 1875. Based on the Edward Turner Bennet edition, but later printings also include the ten letters to Robert Marsham and woodcuts by Bewick.

[12] BELL, THOMAS (ed.), 1876–7. The classic edition in two volumes, containing representative samples of almost everything White wrote, by a man who lived in The Wakes for much of his life.

[13] JEFFERIES, RICHARD (preface), 1887.

[14] ALLEN, GRANT (ed.), 1899. Includes Samuel Taylor Coleridge's manuscript annotations, taken from his copy in the British Museum.

[15] SHARPE, R. BOWDLER (ed.), 1900. Includes an almost complete version of the *Garden Kalendar*.

[16] KEARTON, RICHARD (ed.), 1902. Contains some very fine photographs of Selborne in the late nineteenth century, by Richard and Cherry Kearton.

[17] NICHOLSON, E.M. (ed.), 1929.

[18] MASSINGHAM, H.J. (ed.), 1938.

[19] FISHER, JAMES (ed.), Penguin Books, 1941. Later expanded, and published by the Cresset Press in 1947.

[20] LOCKLEY, R.M. (ed.), Everyman Library, Dent, 1949.

BIBLIOGRAPHY

BIOGRAPHIES, CRITICISM, ETC.

[21] HOLT-WHITE, RASHLEIGH, *The Life and Letters of Gilbert White of Selborne*, John Murray, 1901. The standard biography, by a member of the White family.

[22] HOLT-WHITE, RASHLEIGH, *The Letters to Gilbert White of Selborne from his Intimate Friend and Contemporary The Rev. John Mulso*, R.H. Porter, 1960.

[23] LOCKLEY, R.M., *Gilbert White*, H. F. & G. Witherby, 1954.

[24] NEWTON, ALFRED, entry on Gilbert White in the *Dictionary of National Biography*, Oxford University Press.

[25] RYE, ANTHONY, *Gilbert White and his Selborne*, William Kimber, 1970.

[26] SCOTT, WALTER S., *White of Selborne*, The Falcon Press, 1950.

Short but perceptive critical notes are included in three works on, respectively, the history of the natural history movement, the history of ornithology and the countryside in literature. They serve to place White's discoveries and writing in historical perspective:

[27] ALLEN, DAVID ELLISTON, *The Naturalist in Britain. A Social history*, Allen Lane, 1976.

[28] NICHOLSON, E.M., *Birds in England*, Chapman and Hall, 1926.

[29] WILLIAMS, RAYMOND, *The Country and the City*, Chatto & Windus, 1973.

TOPOGRAPHIES, ETC.

A number of later writers have included accounts of visits to Selborne in the context of larger works. In chronological order again:

[30] COBBETT, WILLIAM, *Rural Rides*, 1830.

[31] HUDSON, W.H., *Birds and Man*, 1902.

[32] HUDSON, W.H., *Hampshire Days*, 1903.

[33] TOMPKINS, H.W., *Selborne*, 1905.

[34] VAUGHAN, JOHN, *The Wild-flowers of Selborne, and other papers*, 1906.

[35] TEALE, EDWIN WAY, *Springtime in Britain*, Cassell, 1970.

[36] BEAMES, IAN, 'Selborne 1774–1974', in *The Bird-Watchers' Book*, edited by John Gooders, David and Charles, 1974.

See also RYE [25].

A NOTE ON THE TEXT AND
ANNOTATIONS

WITH the exception of a few spelling regularizations, the text
follows that of the first edition of the *Natural History*, published
(together with the *Antiquities*) in December 1788 [5]. The first
forty-four letters are to Thomas Pennant, the last sixty-six to
Daines Barrington (though each set is numbered separately).
I have not included the *Antiquities*, nor *Naturalist's Calendar*,
the *Observations on Various Parts of Nature*, or the miscellaneous
poems that are occasionally bound up with it. They have little
in common with the *Natural History* either in quality or in
originality, and do little more than blunt the extraordinary
pitch of condensation which White achieved in his one true
book. I have, however, included full details of these, and
White's other extant writings, in the Bibliography.

White's own notes are indicated by asterisks, etc., and
editorial comments by numerals. It was sadly something of a
tradition for White's early editors to update his observations,
answer all his queries, and use his text as a springboard for their
own theories and dissertations. Edward Turner Bennet's 1837
edition [9] actually appends a note fifteen pages long to a single
remark by White on the treatment of birds in captivity! This
kind of excessive and pedantic pointing-up seems to me not
only a rather arrogant imposition on a work whose sense of
balance and emphasis are finely judged, but against the very
spirit of what White was trying to achieve. His chief concern,
and the essence of his originality as a naturalist, was his belief in
first-hand observation, and his rejection of received opinions.
I have tried therefore to keep the editorial notes to a minimum,
and to confine them to clarification rather than comment. I
have clarified the identity of species where there has been a

change of English name (but have not tried to regularize White's variable scientific Latin, provided it is clear what he is describing); and the identity, too, of his correspondents and references. I have also pointed out the few instances where he is now known to have been in error, and the grosser changes in the natural history of the area (e.g. where a bird mentioned by White has become extinct in England).

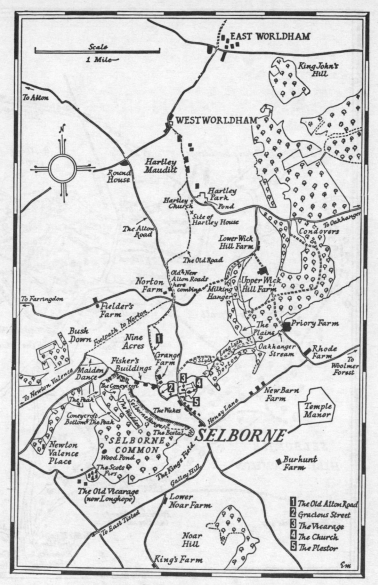

Map of Selborne and its environs

Map of Selborne village

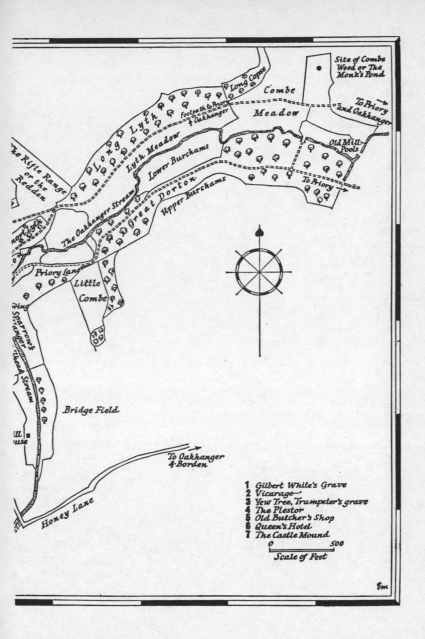

Site of Combe
Wood or The
Monk's Pond

To Priory
and Oakhanger

Long Copse
Combe
Meadow
Footpath to Priory
& Oakhanger
Luth Meadow
Old Mill
Pools
Long Lyth
Lower Burchams
Great Dorton
To Priory
The Rifle Range
or the
Redder
Upper Burchams
The Oakhanger Stream
Priory Lane
Little
Combe
Sparrow's
Stranger
...end Stream
Bridge Field
...ill
...use
To Oakhanger
4 Borden

Honey Lane

1 Gilbert White's Grave
2 Vicarage
3 Yew Tree, Trumpeter's grave
4 The Plestor
5 Old Butcher's Shop
6 Queen's Hotel
7 The Castle Mound

0   500
Scale of Feet

Em.

THE
NATURAL HISTORY
AND
ANTIQUITIES
OF
SELBORNE,
IN THE
COUNTY of SOUTHAMPTON:

WITH

ENGRAVINGS, AND AN APPENDIX.

— — — " ego Apis Matinæ
" More modoque
" Grata carpentis — — — per laborem
" Plurimum," — — — — — HOR.

" Omnia benè defcribere, quæ in hoc mundo, a Deo facta, aut Naturæ creatæ viribus
" elaborata fuerunt, opus eft non unius hominis, nec unius ævi.   Hinc *Fauna* & *Flora*
" utiliffimæ; hinc *Monographi* præftantiffimi."   SCOPOLI ANN. HIST. NAT.

LONDON:
PRINTED BY T. BENSLEY;
FOR B. WHITE AND SON, AT HORACE'S HEAD, FLEET STREET.
M,DCC,LXXXIX.

On previous page: *Facsimile of the title-page of the first edition,
published in December 1788 (though the date of publication
on the title-page is 1789).*

ADVERTISEMENT

THE Author of the following Letters takes the liberty, with all proper deference, of laying before the public his idea of *parochial history*, which, he thinks, ought to consist of natural productions and occurrences as well as antiquities. He is also of opinion that if stationary men would pay some attention to the districts on which they reside, and would publish their thoughts respecting the objects that surround them, from such materials might be drawn the most complete county-histories, which are still wanting in several parts of this kingdom, and in particular in the county of Southampton.

And here he seizes the first opportunity, though a late one, of returning his most grateful acknowledgments to the reverend the President and the reverend and worthy the Fellows of Magdalen College in the University of Oxford, for their liberal behaviour in permitting their archives to be searched by a member of their own society, so far as the evidences therein contained might respect the parish and priory of Selborne. To that gentleman also, and his assistant, whose labours and attention could only be equalled by the very kind manner in which they were bestowed, many and great obligations are also due.

Of the authenticity of the documents above-mentioned there can be no doubt, since they consist of the identical deeds and records that were removed to the College from the Priory at the time of its dissolution; and, being carefully copied on the spot, may be depended on as genuine; and, never having been made public before, may gratify the curiosity of the antiquary, as well as establish the credit of the history.

If the writer should at all appear to have induced any of his readers to pay a more ready attention to the wonders of the Creation, too frequently overlooked as common occurrences; or if he should by any means, through his researches, have lent

3

an helping hand towards the enlargement of the boundaries of historical and topographical knowledge; or if he should have thrown some small light upon ancient customs and manners, and especially on those that were monastic, his purpose will be fully answered. But if he should not have been successful in any of these his intentions, yet there remains this consolation behind – that these his pursuits, by keeping the body and mind employed, have, under Providence, contributed to much health and cheerfulness of spirits, even to old age: and, what still adds to his happiness, have led him to the knowledge of a circle of gentlemen whose intelligent communications, as they have afforded him much pleasing information, so, could he flatter himself with a continuation of them, would they ever be deemed a matter of singular satisfaction and improvement.

GIL: WHITE

Selborne, January 1st, 1788.

Letters to Thomas Pennant

LETTER I[1]

THE parish of Selborne lies in the extreme eastern corner of the county of Hampshire, bordering on the county of Sussex, and not far from the county of Surrey; is about fifty miles south-west of London, in latitude 51, and near midway between the towns of Alton and Petersfield. Being very large and extensive, it abuts on twelve parishes, two of which are in Sussex, viz., Trotton and Rogate. If you begin from the south and proceed westward, the adjacent parishes are Emshot, Newton Valance, Faringdon, Harteley Mauduit, Great Ward le ham, Kingsley, Hedleigh, Bramshot, Trotton, Rogate, Lysse, and Greatham. The soils of this district are almost as various and diversified as the views and aspects. The high part to the south-west consists of a vast hill of chalk, rising three hundred feet above the village; and is divided into a sheep down, the high wood, and a long hanging wood called the Hanger. The covert of this eminence is altogether beech, the most lovely of all forest trees, whether we consider its smooth rind or bark, its glossy foliage, or graceful pendulous boughs. The down, or sheep-walk, is a pleasing park-like spot, of about one mile by half that space, jutting out on the verge of the hill-country, where it begins to break down into the plains, and commanding a very engaging view, being an assemblage of hill, dale, wood-lands, heath, and water. The prospect is bounded to the south-east and east by the vast range of mountains called the Sussex-downs, by Guild-down near Guildford, and by the Downs round Dorking, and Ryegate in Surrey, to the north-east, which altogether, with the country beyond Alton and Farnham, form a noble and extensive outline.

At the foot of this hill, one stage or step from the uplands, lies the village, which consists of one single straggling street,

7

three-quarters of a mile in length, in a sheltered vale, and running parallel with the Hanger. The houses are divided from the hill by a vein of stiff clay (good wheat-land), yet stand on a rock of white stone, little in appearance removed from chalk; but seems so far from being calcareous, that it endures extreme heat. Yet that the freestone still preserves somewhat that is analogous to chalk, is plain from the beeches which descend as low as those rocks extend, and no farther, and thrive as well on them, where the ground is steep, as on the chalks.

The cart-way of the village divides, in a remarkable manner, two very incongruous soils. To the south-west is a rank-clay, that requires the labour of years to render it mellow; while the gardens to the north-east, and small enclosures behind, consist of a warm, forward, crumbling mould, called *black malm*, which seems highly saturated with vegetable and animal manure; and these may perhaps have been the original site of the town; while the wood and coverts might extend down to the opposite bank.

At each end of the village, which runs from south-east to north-west, arises a small rivulet: that at the north-west end frequently fails; but the other is a fine perennial spring, little influenced by drought or wet seasons, called Well-head.* This breaks out of some high grounds joining to Nore Hill, a noble chalk promontory, remarkable for sending forth two streams into two different seas. The one to the south becomes a branch of the Arun, running to Arundel, and so falling into the British Channel: the other to the north. The Selborne stream makes one branch of the Wey; and meeting the Black-down stream at Hedleigh, and the Alton and Farnham stream at Tilford-

---

*This spring produced, September 14, 1781, after a severe hot summer, and a preceding dry spring and winter, nine gallons of water in a minute, which is five hundred and forty in an hour, and twelve thousand nine hundred and sixty, or two hundred and sixteen hogsheads, in twenty-four hours, or one natural day. At this time many of the wells failed, and all the ponds in the vales were dry.

bridge, swells into a considerable river, navigable at Godalming from whence it passes to Guildford, and so into the Thames at Weybridge; and thus at the Nore into the German Ocean.

Our wells, at an average, run to about sixty-three feet, and when sunk to that depth seldom fail; but produce a fine limpid water, soft to the taste, and much commended by those who drink the pure element, but which does not lather well with soap.

To the north-west, north and east of the village, is a range of fair enclosures, consisting of what is called a *white malm*, a sort of rotten or rubble stone, which, when turned up to the frost and rain, moulders to pieces, and becomes manure to itself.*

Still on to the north-east, and a step lower, is a kind of white land, neither chalk nor clay, neither fit for pasture nor for the plough, yet kindly for hops, which root deep into the freestone, and have their poles and wood for charcoal growing just at hand. This white soil produces the brightest hops.

As the parish still inclines down towards Wolmer-forest, at the juncture of the clays and sand the soil becomes a wet, sandy loam, remarkable for timber, and infamous for roads. The oaks of Temple and Blackmoor stand high in the estimation of purveyors, and have furnished much naval timber; while the trees on the freestone grow large, but are what workmen call *shakey*, and so brittle as often to fall to pieces in sawing. Beyond the sandy loam the soil becomes an hungry lean sand, till it mingles with the forest; and will produce little without the assistance of lime and turnips.[2]

*This soil produces good wheat and clover.

LETTER II

IN the court of Norton-farmhouse, a manor farm to the north-west of the village, on the white malms, stood within these twenty years a broad-leaved elm, or wych hazel,[3] *ulmus folio latissimo scabro* of Ray, which, though it had lost a considerable leading bough in the great storm in the year 1703, equal to a moderate tree, yet, when felled, contained eight loads of timber; and, being too bulky for a carriage, was sawn off at seven feet above the butt, where it measured near eight feet in the diameter. This elm I mention to show to what a bulk planted elms may attain; as this tree must certainly have been such from its situation.

In the centre of the village, and near the church, is a square piece of ground surrounded by houses, and vulgarly called the Plestor. In the midst of this spot stood, in old times, a vast oak, with a short squat body, and huge horizontal arms extending almost to the extremity of the area. This venerable tree, surrounded with stone steps, and seats above them, was the delight of old and young, and a place of much resort in summer evenings; where the former sat in grave debate, while the latter frolicked and danced before them. Long might it have stood, had not the amazing tempest in 1703 overturned it at once, to the infinite regret of the inhabitants, and the vicar, who bestowed several pounds in setting it in its place again; but all his care could not avail; the tree sprouted for a time, then withered and died. This oak I mention to show to what a bulk planted oaks also may arrive: and planted this tree must certainly have been, as will appear from what will be said farther concerning this area, when we enter on the antiquities of Selborne.

On the Blackmoor estate there is a small wood called Losel's, of a few acres, that was lately furnished with a set of oaks of a peculiar growth and great value; they were tall and taper like firs, but standing near together had very small heads, only a little brush without any large limbs. About twenty years ago the bridge at the Toy, near Hampton-court, being much decayed, some trees were wanted for the repairs that were fifty feet long without bough, and would measure twelve inches diameter at the little end. Twenty such trees did a purveyor find in this little wood, with this advantage, that many of them answered the description at sixty feet. These trees were sold for twenty pounds apiece.

In the centre of this grove there stood an oak, which, though shapely and tall on the whole, bulged out into a large excrescence about the middle of the stem. On this a pair of ravens had fixed their residence for such a series of years, that the oak was distinguished by the title of the Raven-tree.[4] Many were the attempts of the neighbouring youths to get at this eyry: the difficulty whetted their inclinations, and each was ambitious of surmounting the arduous task. But, when they arrived at the swelling, it jutted out so in their way, and was so far beyond their grasp, that the most daring lads were awed, and acknowledged the undertaking to be too hazardous. So the ravens built on, nest upon nest, in perfect security, till the fatal day arrived in which the wood was to be levelled. It was in the month of February, when those birds usually sit. The saw was applied to the butt, the wedges were inserted into the opening, the woods echoed to the heavy blows of the beetle or mallet, the tree nodded to its fall; but still the dam sat on. At last, when it gave way, the bird was flung from her nest; and, though her parental affection deserved a better fate, was whipped down by the twigs, which brought her dead to the ground.

LETTER III

THE fossil-shells of this district, and sorts of stone, such as have fallen within my observation, must not be passed over in silence. And first I must mention, as a great curiosity, a specimen that was ploughed up in the chalky fields, near the side of the down, and given to me for the singularity of its appearance, which, to an incurious eye, seems like a petrified fish of about four inches long, the *cardo* passing for an head and mouth. It is in reality a bivalve of the Linnæan genus of *Mytilus*, and the species of *Crista Galli*; called by Lister, *Rastellum*; by Rumphius, *Ostreum plicatum minus*; by D'Argenville, *Auris Porci*, s. *Crista Galli*, and by those who make collections *cock's comb*.[5] Though I applied to several such in London, I could never meet with an entire specimen; nor could I ever find in books any engraving from a perfect one. In the superb museum at Leicester-house,[6] permission was given me to examine for this article; and though I was disappointed as to the fossil, I was highly gratified with the sight of several of the shells themselves in high preservation. This bivalve is only known to inhabit the Indian Ocean, where it fixes itself to a zoophyte, known by the name *Gorgonia*. The curious foldings of the suture, the one into the other, the alternate flutings or grooves, and the curved form of my specimen being much easier expressed by the pencil than by words, I have caused it to be drawn and engraved.

Cornua Ammonis[7] are very common about this village. As we were cutting an inclining path up the Hanger, the labourers found them frequently on that steep, just under the soil, in the chalk, and of a considerable size. In the lane above Well-head, in the way to Emshot, they abound in the bank, in a darkish sort of marl; and are usually very small and soft: but in Clay's Pond, a little farther on, at the end of the pit, where the soil is dug out for manure, I have occasionally observed them of large dimensions, perhaps fourteen or sixteen inches in diameter. But as these did not consist of firm stone, but were formed of a kind of *terra lapidosa*, or hardened clay, as soon as they were exposed to the rains and frost they mouldered away. These seemed as if they were a very recent production. In the chalk-pit, at the north-west end of the Hanger, large *nautili* are sometimes observed.

In the very thickest strata of our freestone, and at considerable depths, well-diggers often find large scallops or *pectines*, having both shells deeply striated, and ridged and furrowed alternately. They are highly impregnated with, if not wholly composed of, the stone of the quarry.

LETTER IV

As in a former letter the freestone[8] of this place has been only mentioned incidentally, I shall here become more particular.

This stone is in great request for hearth-stones and the beds of ovens: and in lining of lime-kilns it turns to good account; for the workmen use sandy loam instead of mortar; the sand of which fluxes★ and runs by the intense heat, and so cases over the whole face of the kiln with a strong vitrified coat like glass, that it is well preserved from injuries of weather, and endures thirty or forty years. When chiseled smooth, it makes elegant fronts for houses, equal in colour and grain to the Bath stone; and superior in one respect, that, when seasoned, it does not scale. Decent chimney-pieces are worked from it of much closer and finer grain than Portland; and rooms are floored

★There may probably be also in the chalk itself that is burnt for lime a proportion of sand: for few chalks are so pure as to have none.

with it; but it proves rather too soft for this purpose. It is a freestone, cutting in all directions; yet has something of a grain parallel with the horizon, and therefore should not be surbedded, but laid in the same position as it grows in the quarry.* On the ground abroad this firestone will not succeed for pavements, because, probably, some degree of saltness prevailing within it, the rain tears the slabs to pieces.† Though this stone is too hard to be acted on by vinegar, yet both the white part, and even the blue rag, ferments strongly in mineral acids. Though the white stone will not bear wet, yet in every quarry at intervals there are thin strata of blue rag, which resist rain and frost; and are excellent for pitching of stables, paths, and courts, and for building of dry walls against banks, a valuable species of fencing, much in use in this village, and for mending of roads. This rag is rugged and stubborn, and will not hew to a smooth face; but is very durable: yet, as these strata are shallow and lie deep, large quantities cannot be procured but at considerable expense. Among the blue rags turn up some blocks tinged with a stain of yellow or rust colour, which seem to be nearly as lasting as the blue; and every now and then balls of a friable substance, like rust of iron, called rust balls.

In Wolmer-forest I see but one sort of stone, called by the workmen sand, or forest-stone. This is generally of the colour of rusty iron, and might probably be worked as iron ore; is very hard and heavy, and of a firm, compact texture, and composed of a small roundish crystalline grit, cemented together by a brown, terrene, ferruginous matter; will not cut without difficulty, nor easily strike fire with steel. Being often found in

---

*To *surbed* stone is to set it edgewise, contrary to the posture it had in the quarry, says Dr Plot, *Oxfordsh.*, p. 77. But surbedding does not succeed in our dry walls; neither do we use it so in ovens, though he says it is best for Teynton stone.

†'Firestone is full of salts, and has no sulphur: must be close grained, and have no interstices. Nothing supports fire like salts; saltstone perishes exposed to wet and frost.' Plot's *Staff.*, p. 152.

broad flat pieces, it makes good pavement for paths about houses, never becoming slippery in frost or rain; is excellent for dry walls, and is sometimes used in buildings. In many parts of that waste it lies scattered on the surface of the ground; but is dug on Weaver's-down, a vast hill on the eastern verge of that forest, where the pits are shallow, and the stratum thin. This stone is imperishable.

From a notion of rendering their work the more elegant, and giving it a finish, masons chip this stone into small fragments about the size of the head of a large nail; and then stick the pieces into the wet mortar along the joints of their freestone walls: this embellishment carries an odd appearance, and has occasioned strangers sometimes to ask us pleasantly, 'whether we fastened our walls together with tenpenny nails.'

LETTER V

AMONG the singularities of this place the two rocky hollow lanes, the one to Alton, and the other to the forest, deserve our attention. These roads, running through the malm lands, are, by the traffic of ages, and the fretting of water, worn down through the first stratum of our freestone, and partly through the second; so that they look more like water-courses than roads; and are bedded with naked rag for furlongs together. In many places they are reduced sixteen or eighteen feet beneath the level of the fields; and after floods, and in frosts, exhibit very grotesque and wild appearances, from the tangled roots that are twisted among the strata, and from the torrents rushing down their broken sides; and especially when those cascades are frozen into icicles, hanging in all the fanciful shapes of frost-work. These rugged gloomy scenes affright the ladies when they peep down into them from the paths above, and make timid horsemen shudder while they ride along them; but delight the naturalist with their various botany, and particularly with their curious *filices*9 with which they abound.

The manor of Selborne, was it strictly looked after, with its kindly aspects, and all its sloping coverts, would swarm with game; even now hares, partridges, and pheasants abound; and in old days woodcocks were as plentiful. There are few quails, because they more affect open fields than enclosures; after harvest some few landrails are seen.10

The parish of Selborne, by taking in so much of the forest, is a vast district. Those who tread the bounds are employed part of three days in the business, and are of opinion that the outline, in all its curves and indentings, does not comprise less than thirty miles.

The village stands in a sheltered spot, secured by the Hanger from the strong westerly winds. The air is soft, but rather moist from the effluvia of so many trees; yet perfectly healthy and free from agues.

The quantity of rain that falls on it is very considerable, as may be supposed in so woody and mountainous a district. As my experience in measuring the water is but of short date, I am not qualified to give the mean quantity.*

I only know that:

	Inch.	Hund.
From May 1, 1779, the end of the year, there fell .	28	37!
From Jan. 1, 1780, to Jan. 1, 1781, there fell . .	27	32
From Jan. 1, 1781, to Jan. 1, 1782, there fell . .	30	71
From Jan. 1, 1782, to Jan. 1, 1783, there fell . .	50	26!
From Jan. 1, 1783, to Jan. 1, 1784, there fell . .	33	71
From Jan. 1, 1784, to Jan. 1, 1785, there fell . .	33	80
From Jan. 1, 1785, to Jan. 1, 1786, there fell . .	31	55
From Jan. 1, 1786, to Jan. 1, 1787, there fell . .	39	57

*A very intelligent gentleman assures me (and he speaks from upwards of forty years' experience) that the mean rain of any place cannot be ascertained till a person has measured it for a very long period. 'If I had only measured the rain,' says he, 'for the four first years, from 1740 to 1743, I should have said the mean rain at Lyndon was 16½ inches for

The village of Selborne, and large hamlet of Oakhanger, with the single farms, and many scattered houses along the verge of the forest, contain upwards of six hundred and seventy inhabitants.† We abound with poor; many of whom are sober and industrious, and live comfortably in good stone or brick cottages, which are glazed, and have chambers above stairs: mud buildings we have none. Besides the employment from husbandry the men work in hop gardens, of which we have many; and fell and bark timber. In the spring and summer the women weed the corn; and enjoy a second harvest in September by hop-picking. Formerly, in the dead months they availed themselves greatly by spinning wool, for making of *barragons*, a genteel corded stuff, much in vogue at that time for summer wear; and chiefly manufactured at Alton, a neighbouring town, by some of the people called Quakers: but from circumstances

---

the year; if from 1740 to 1750, 18½ inches. The mean rain before 1763 was 20¼; from 1763 and since, 25½; from 1770 to 1780, 26. If only 1773, 1774 and 1775 had been measured, Lyndon mean rain would have been called 32 inches.'

†*A STATE of the Parish of SELBORNE, taken October 4, 1783.*

The number of tenements or families, 136.

The number of inhabitants in the street is . . . 313		Total, 676; near five inhabitants to each tenement.
In the rest of the parish . 363		

In the time of the Rev. Gilbert White, vicar, who died in 1727–8, the number of inhabitants was computed at about 500.

Average of baptisms for 60 years.

From 1720 to 1729, both years inclusive .	Males 6,9 Females 6,0	12,9
From 1730 to 1739, both years inclusive .	Males 8,2 Females 7,1	15,3
From 1740 to 1749, inclusive . .	Males 9,2 Females 6,6	15,8
From 1750 to 1759, inclusive . .	Males 7,6 Females 8,1	15,7

this trade is at an end.* The inhabitants enjoy a good share of health and longevity: and the parish swarms with children.

From 1760 to 1769, inclusive . . { Males 9,1 } 18,0
{ Females 8,9 }

From 1770 to 1779, inclusive . . { Males 10,5 } 20,3
{ Females 9,8 }

Total baptisms of Males 515 } 980
Females 465 }

Total of baptisms from 1720 to 1779,
both inclusive, 60 years . . . 980.

Average of burials for 60 years.

From 1720 to 1729, both years inclusive . { Males 4,8 } 9,9
{ Females 5,1 }

From 1730 to 1739, both years inclusive . { Males 4,8 } 10,6
{ Females 5,8 }

From 1740 to 1749, inclusive . . { Males 4,6 } 8,4
{ Females 3,8 }

From 1750 to 1759, inclusive . . { Males 4,9 } 10,0
{ Females 5,1 }

From 1760 to 1769, inclusive . . { Males 6,9 } 13,4
{ Females 6,5 }

From 1770 to 1779, inclusive . . { Males 5,5 } 11,7
{ Females 6,2 }

Total of burials of Males 315 } 640
Females 325 }

Total of burials from 1720 to 1779
both inclusive, 60 years . 640.

Baptisms exceed burials by more than one-third.
Baptisms of Males exceed Females by one-tenth, or one in ten.
Burials of Females exceed Males by one in thirty.
It appears that a child, born and bred in this parish, has an equal
chance to live above forty years.
Twins thirteen times, many of whom dying young have lessened
the chance for life.
Chances for life in men and women appear to be equal.

*Since the passage above was written, I am happy in being able to say that the spinning employment is a little revived, to the no small comfort of the industrious housewife.

A TABLE *of the* Baptisms, Burials, *and* Marriages, *from* January 2, 1761, *to* December 25, 1780, *in the* Parish *of* SELBORNE.

	BAPTISMS			BURIALS			MAR.
	Males.	Females.	Total.	Males.	Females.	Total.	
1761	8	10	18	2	4	6	3
1762	7	8	15	10	10	20	6
1763	8	10	18	3	4	7	5
1764	11	9	20	10	8	18	6
1765	12	6	18	9	7	16	6
1766	9	13	22	10	6	16	4
1767	14	5	19	6	5	11	2
1768	7	6	13	2	5	7	6
1769	9	14	23	6	5	11	2
1770	10	13	23	4	7	11	3
1771	10	6	16	3	4	7	4
1772	11	10	21	6	10	16	3
1773	8	5	13	7	5	12	3
1774	6	13	19	2	8	10	1
1775	20	7	27	13	8	21	6
1776	11	10	21	4	6	10	6
1777	8	13	21	7	2	9	4
1778	7	13	20	3	9	12	5
1779	14	8	22	5	6	11	5
1780	8	9	17	11	4	15	3
	198	188	386	123	123	246	83

During this period of twenty years the births of Males exceeded those of Females . . 10.

The burials of each sex were equal.

And the births exceeded the deaths . . 140.

LETTER VI

SHOULD I omit to describe with some exactness the forest of Wolmer, of which three-fifths perhaps lie in this parish, my account of Selborne would be very imperfect, as it is a district abounding with many curious productions, both animal and vegetable; and has often afforded me much entertainment both as a sportsman and as a naturalist.

The royal forest of Wolmer is a tract of land of about seven miles in length, by two and a half in breadth, running nearly from north to south, and is abutted on, to begin to the south, and so to proceed eastward, by the parishes of Greatham, Lysse, Rogate, and Trotton, in the county of Sussex; by Bramshot, Hedleigh, and Kingsley. This royalty consists entirely of sand covered with heath and fern; but is somewhat diversified with hills and dales, without having one standing tree in the whole extent.[11] In the bottoms, where the waters stagnate, are many bogs, which formerly abounded with subterraneous trees; though Dr Plot[12] says positively,* that 'there never were any fallen trees hidden in the mosses of the southern counties.' But he was mistaken: for I myself have seen cottages on the verge of this wild district, whose timbers consisted of a black hard wood, looking like oak, which the owners assured me they procured from the bogs by probing the soil with spits, or some such instruments:[13] but the peat is so much cut out, and the moors have been so well examined, that none has been found of late.† Besides the oak, I have also been shown pieces of fossil-

* See his *Hist.* of *Staffordshire*.
†Old people have assured me, that on a winter's morning they have discovered these trees in the bogs, by the hoar frost, which lay longer over the space where they were concealed, than on the surrounding morass. Nor does this seem to be a fanciful notion, but consistent with

wood of a paler colour, and softer nature, which the inhabitants called fir: but, upon a nice examination, and trial by fire, I could discover nothing resinous in them; and therefore rather suppose that they were parts of a willow or alder, or some such aquatic tree.

This lonely domain is a very agreeable haunt for many sorts of wild fowls, which not only frequent it in the winter, but breed there in the summer; such as lapwings, snipes, wild-ducks, and, as I have discovered within these few years, teals. Partridges in vast plenty are bred in good seasons on the verge of this forest, into which they love to make excursions: and in particular, in the dry summer of 1740 and 1741, and some years after, they swarmed to such a degree, that parties of unreasonable sportsmen killed twenty and sometimes thirty brace in a day.

But there was a nobler species of game in this forest, now extinct, which I have heard old people say abounded much before shooting flying became so common, and that was the heath-cock, black-game, or grouse.[15] When I was a little boy I recollect one coming now and then to my father's table. The

---

true philosophy. Dr Hales saith, 'That the warmth of the earth, at some depth under ground, has an influence in promoting a thaw, as well as the change of the weather from a freezing to a thawing state, is manifest, from this observation, viz. Nov. 29, 1731, a little snow having fallen in the night, it was, by eleven the next morning, mostly melted away on the surface of the earth, except in several places in Bushy Park, where there were drains dug and covered with earth, on which the snow continued to lie, whether those drains were full of water or dry; as also where elm-pipes lay under ground: a plain proof this, that those drains intercepted the warmth of the earth from ascending from greater depths below them: for the snow lay where the drain had more than four feet depth of earth over it. It continued also to lie on thatch, tiles, and the tops of walls.' See Hales's *Hæmastatics*,[14] p. 360. *Quære.* – Might not such observations be reduced to domestic use, by promoting the discovery of old obliterated drains and wells about houses; and in Roman stations and camps lead to the finding of pavements, baths and graves, and other hidden relics of curious antiquity?

last pack remembered was killed about thirty-five years ago; and within these ten years one solitary greyhen was sprung by some beagles in beating for a hare. The sportsmen cried out, 'A hen pheasant'; but a gentleman present, who had often seen grouse in the north of England, assured me that it was a greyhen.

Nor does the loss of our black game prove the only gap in the *Fauna Selborniensis*; for another beautiful link in the chain of beings is wanting, I mean the red deer, which toward the beginning of this century amounted to about five hundred head, and made a stately appearance. There is an old keeper, now alive, named Adams, whose great-grandfather (mentioned in a perambulation taken in 1635), grandfather, father, and self, enjoyed the head keepership of Wolmer-forest in succession for more than an hundred years. This person assures me, that his father has often told him, that Queen Anne, as she was journeying on the Portsmouth road, did not think the forest of Wolmer beneath her royal regard. For she came out of the great road at Lippock, which is just by, and reposing herself on a bank smoothed for that purpose, lying about half a mile to the east of Wolmer-pond, and still called Queen's-bank, saw with great complacency and satisfaction the whole herd of red deer brought by the keepers along the vale before her, consisting then of about five hundred head. A sight this, worthy the attention of the greatest sovereign! But he further adds that, by means of the Waltham blacks,[16] or, to use his own expression, as soon as they began *blacking*, they were reduced to about fifty head, and so continued decreasing till the time of the late Duke of Cumberland. It is now more than thirty years ago that his highness sent down an huntsman, and six yeoman-prickers, in scarlet jackets laced with gold, attended by the stag-hounds; ordering them to take every deer in this forest alive, and convey them in carts to Windsor. In the course of the summer they caught every stag, some of which showed extraordinary diversion; but, in the following winter, when the hinds were also carried off, such fine chases were exhibited as served the country people for matter of talk and wonder for years afterwards. I saw myself one of the yeoman-prickers single out a stag from the herd, and must confess that it was the most curious feat of activity I ever beheld, superior to anything in Mr Astley's riding-school. The exertions made by the horse and deer much exceeded all my expectations; though the former greatly excelled the latter in speed. When the devoted deer was separated from his companions, they gave him, by their watches, law, as they called it, for twenty minutes; when, sounding their horns, the stop-dogs were permitted to pursue, and a most gallant scene ensued.

LETTER VII

THOUGH large herds of deer do much harm to the neighbourhood, yet the injury to the morals of the people is of more moment than the loss of their crops. The temptation is irresistible; for most men are sportsmen by constitution: and there is such an inherent spirit for hunting in human nature, as scarce any inhibitions can restrain. Hence, towards the beginning of this century, all this country was wild about deer-stealing. Unless he was a *hunter*, as they affected to call themselves, no young person was allowed to be possessed of manhood or gallantry. The Waltham blacks at length committed such enormities, that government was forced to interfere with that severe and sanguinary act called the Black Act,* which now comprehends more felonies than any law that ever was framed before. And, therefore, a late bishop of Winchester, when urged to re-stock Waltham-chase,† refused, from a motive

* Statute 9 Geo. I. c. 22.
† This chase remains unstocked to this day; the bishop was Dr Hoadly.

worthy of a prelate, replying that 'it had done mischief enough already.'

Our old race of deer-stealers are hardly extinct yet: it was but a little while ago that, over their ale, they used to recount the exploits of their youth; such as watching the pregnant hind to her lair, and, when the calf was dropped, paring its feet with a penknife to the quick to prevent its escape, till it was large and fat enough to be killed; the shooting at one of their neighbours with a bullet in a turnip-field by moonshine, mistaking him for a deer; and the losing a dog in the following extraordinary manner: Some fellows, suspecting that a calf new-fallen was deposited in a certain spot of thick fern, went, with a lurcher, to surprise it; when the parent hind rushed out of the brake, and, taking a vast spring with all her feet close together, pitched upon the neck of the dog, and broke it short in two.

Another temptation to idleness and sporting was a number of rabbits, which possessed all the hillocks and dry places: but these being inconvenient to the huntsmen, on account of their burrows, when they came to take away the deer, they permitted the country people to destroy them all.

Such forests and wastes, when their allurements to irregularities are removed, are of considerable service to neighbourhoods that verge upon them, by furnishing them with peat and turf for their firing; with fuel for the burning their lime; and with ashes for their grasses; and by maintaining their geese and their stock of young cattle at little or no expense.

The manor farm of the parish of Greatham has an admitted claim, I see (by an old record taken from the Tower of London), of turning all live stock on the forest at proper seasons, *bidentibus exceptis.** The reason, I presume, why sheep† are

---

* For the privilege the owner of that estate used to pay to the king annually seven bushels of oats.

† In the Holt, where a full stock of fallow-deer has been kept up till lately, no sheep are admitted to this day.

excluded, is, because, being such close grazers, they would pick out all the finest grasses, and hinder the deer from thriving.

Though (by statute 4 and 5 W. and Mary, c. 23) 'to burn on any waste, between Candlemas and Midsummer, any grig, ling, heath and furze, goss or fern, is punishable with whipping and confinement in the house of correction'; yet, in this forest, about March or April, according to the dryness of the season, such vast heath-fires are lighted up, that they often get to a masterless head, and, catching the hedges, have sometimes been communicated to the underwoods, woods, and coppices, where great damage has ensued. The plea for these burnings is, that, when the old coat of heath, etc., is consumed, young will sprout up, and afford much tender browse for cattle; but, where there is large old furze, the fire, following the roots, consumes the very ground; so that for hundreds of acres nothing is to be seen but smother and desolation, the whole circuit round looking like the cinders of a volcano; and the soil being quite exhausted, no traces of vegetation are to be found for years. These conflagrations, as they take place usually with a north-east or east wind, much annoy this village with their smoke, and often alarm the country; and, once in particular, I remember that a gentleman, who lives beyond Andover, coming to my house, when he got on the downs between that town and Winchester, at twenty-five miles distance, was surprised much with smoke and a hot smell of fire; and concluded that Alresford was in flames; but, when he came to that town, he then had apprehensions for the next village, and so on to the end of his journey.

On two of the most conspicuous eminences of this forest, stand two arbours or bowers, made of the boughs of oaks; the one called Waldon-lodge, the other Brimstone-lodge: these the keepers renew annually on the feast of St Barnabas,[17] taking the old materials for a perquisite. The farm called Blackmoor, in this parish, is obliged to find the posts and brush-wood for the former; while the farms at Greatham, in rotation, furnish for the latter; and are all enjoined to cut and deliver the materials at the spot. This custom I mention, because I look upon it to be of very remote antiquity.

LETTER VIII

O n the verge of the forest, as it is now circumscribed, are three considerable lakes, two in Oakhanger, of which I have nothing particular to say; and one called Bin's or Bean's Pond, which is worthy the attention of a naturalist or a sportsman. For, being crowded at the upper end with willows, and with the *carex cespitosa,*[18] it affords such a safe and pleasing shelter to wild-ducks, teals, snipes, etc., that they breed there. In the winter this covert is also frequented by foxes, and sometimes by pheasants; and the bogs produce many curious plants. [For which consult Letter XLI to Mr Barrington.]

By a perambulation of Wolmer-forest and the Holt, made in 1635, and in the eleventh year of Charles the First (which now lies before me), it appears that the limits of the former are much circumscribed. For, to say nothing on the farther side, with which I am not so well acquainted, the bounds on this side, in old times, came into Binswood; and extended to the ditch of Ward le ham park, in which stands the curious mount called King John's Hill, and Lodge Hill; and to the verge of Hartley Mauduit, called Mauduit-hatch; comprehending also Short-heath, Oakhanger, and Oakwoods; a large district, now private property, though once belonging to the royal domain.

*I mean that sort which, rising into tall hassocks, is called by the foresters *torrets*; a corruption, I suppose, of turrets.

Note. In the beginning of the summer 1787 the royal forests of Wolmer and Holt were measured by persons sent down by government.

It is remarkable that the term *purlieu*[19] is never once mentioned in this long roll of parchment. It contains, besides the perambulation, a rough estimate of the value of the timbers, which were considerable, growing at that time in the district of the Holt; and enumerates the officers, superior and inferior, of those joint forests, for the time being, and their ostensible fees and perquisites. In those days, as at present, there were hardly any trees in Wolmer-forest.

Within the present limits of the forest are three considerable lakes, Hogmer, Cranmer, and Wolmer; all of which are stocked with carp, tench, eels, and perch; but the fish do not thrive well, because the water is hungry, and the bottoms are a naked sand.

A circumstance respecting these ponds, though by no means peculiar to them, I cannot pass over in silence; and that is, that instinct by which in summer all the kine, whether oxen, cows, calves, or heifers, retire constantly to the water during the hotter hours; where, being more exempt from flies, and inhaling the coolness of that element, some belly deep, and some only to mid-leg, they ruminate and solace themselves from about ten in the morning till four in the afternoon, and then return to their feeding. During this great proportion of the day they drop much dung, in which insects nestle; and so supply food for the fish, which would be poorly subsisted but from this contingency. Thus nature, who is a great economist, converts the recreation of one animal to the support of another! Thomson, who was a nice observer of natural occurrences, did not let this pleasing circumstance escape him. He says, in his *Summer*:

> A various group the herds and flocks compose:
> ... on the grassy bank
> Some ruminating lie; while others stand
> Half in the flood, and, often bending, sip
> The circling surface.

Wolmer-pond, so called, I suppose, for eminence sake, is a vast lake for this part of the world, containing, in its whole circumference, 2,646 yards, or very near a mile and a half. The length of the north-west and opposite side is about 704 yards, and the breadth of the south-west end about 456 yards. This measurement, which I caused to be made with good exactness, gives an area of about sixty-six acres, exclusive of a large irregular arm at the north-east corner, which we did not take into the reckoning.

On the face of this expanse of waters, and perfectly secure from fowlers, lie all day long, in the winter season, vast flocks of ducks, teals, and widgeons, of various denominations; where they preen and solace, and rest themselves, till towards sunset, when they issue forth in little parties (for in their natural state they are all birds of the night) to feed in the brooks and meadows; returning again with the dawn of the morning. Had this lake an arm or two more, and were it planted round with thick covert (for now it is perfectly naked), it might make a valuable decoy.

Yet neither its extent, nor the clearness of its water, nor the resort of various and curious fowls, nor its picturesque groups of cattle, can render this *meer* so remarkable as the great quantity of coins that were found in its bed about forty years ago. But, as such discoveries more properly belong to the antiquities of this place, I shall suppress all particulars for the present, till I enter professedly on my series of letters respecting the more remote history of this village and district.

LETTER IX

By way of supplement, I shall trouble you once more on this subject, to inform you that Wolmer, with her sister forest Ayles Holt, *alias* Alice Holt,* as it is called in old records, is held by grant from the crown for a term of years.

The grantees that the author remembers are Brigadier-General Emanuel Scroope Howe, and his lady, Ruperta, who was a natural daughter of Prince Rupert by Margaret Hughs; a Mr Mordaunt, of the Peterborough family, who married a dowager Lady Pembroke; Henry Bilson Legge and lady; and now Lord Stawel, their son.

The lady of General Howe lived to an advanced age, long surviving her husband; and, at her death, left behind her many curious pieces of mechanism of her father's constructing, who was a distinguished mechanic and artist,† as well as warrior; and, among the rest, a very complicated clock, lately in possession of Mr Elmer, the celebrated game-painter at Farnham, in the county of Surrey.

Though these two forests are only parted by a narrow range of enclosures, yet no two soils can be more different: for the Holt consists of a strong loam, of a miry nature, carrying a good turf, and abounding with oaks that grow to be large timber; while Wolmer is nothing but a hungry, sandy, barren waste.

The former, being all in the parish of Binsted, is about two miles in extent from north to south, and near as much from east to west, and contains within it many woodlands and lawns, and

---

*In 'Rot. Inquisit. de statu forest. in Scaccar.,' 36, Ed. 3, it is called *Aisholt*. In the same, 'Tit. *Woolmer* and *Aisholt* Hantisc. Dominus Rex habet unam capellam in *haia* suâ de Kingesle.' '*Haia, sepes, sepimentum, parcus:* a Gall. *haie* and *haye*.' – Spelman's Glossary.

†This prince was the inventor of *mezzotinto*.

the great lodge where the grantees reside; and a smaller lodge, called Goose-green; and is abutted on by the parishes of Kingsley, Frinsham, Farnham, and Bentley; all of which have right of common.

One thing is remarkable; that, though the Holt has been of old well-stocked with fallow-deer, unrestrained by any pales or fences more than a common hedge, yet they were never seen within the limits of Wolmer; nor were the red deer of Wolmer ever known to haunt the thickets or glades of the Holt.

At present the deer of the Holt are much thinned and reduced by the night-hunters, who perpetually harass them in spite of the efforts of numerous keepers, and the severe penalties that have been put in force against them as often as they have been detected, and rendered liable to the lash of the law. Neither fines nor imprisonment can deter them: so impossible is it to extinguish the spirit of sporting, which seems to be inherent in human nature.

General Howe turned out some German wild boars and sows in his forests, to the great terror of the neighbourhood; and, at one time, a wild bull or buffalo: but the country rose upon them and destroyed them.

A very large fall of timber, consisting of about one thousand oaks, has been cut this spring (viz., 1784) in the Holt forest; one-fifth of which, it is said, belongs to the grantee, Lord Stawel. He lays claim also to the lop and top:[20] but the poor of the parishes of Binsted and Frinsham, Bentley and Kingsley, assert that it belongs to them; and, assembling in a riotous manner, have actually taken it all away. One man, who keeps a team, has carried home, for his share, forty stacks of wood. Forty-five of these people his lordship has served with actions. These trees, which were very sound and in high perfection, were winter-cut, viz., in February and March, before the bark would run. In old times the Holt was estimated to be eighteen miles, computed measure, from water-carriage, viz., from the town of Chertsey, on the Thames; but now it is not half that distance, since the Wey is made navigable up to the town of Godalming in the county of Surrey.

LETTER X

August 4, 1767.

It has been my misfortune never to have had any neighbours whose studies have led them towards the pursuit of natural knowledge; so that, for want of a companion to quicken my industry and sharpen my attention, I have made but slender progress in a kind of information to which I have been attached from my childhood.

As to swallows (*hirundines rusticæ*) being found in a torpid state during the winter in the Isle of Wight, or any part of this country, I never heard any such account worth attending to.[21] But a clergyman, of an inquisitive turn, assures me that, when he was a great boy, some workmen, in pulling down the battle-ments of a church tower early in the spring, found two or three swifts (*hirundines apodes*) among the rubbish, which were, at first appearance, dead, but, on being carried toward the fire, revived. He told me that, out of his great care to preserve them, he put them in a paper bag, and hung them by the kitchen fire, where they were suffocated.

Another intelligent person has informed me that, while he was a schoolboy at Brighthelmstone, in Sussex, a great frag-ment of the chalk cliff fell down one stormy winter on the beach; and that many people found swallows among the rubbish; but, on my questioning him whether he saw any of those birds himself, to my no small disappointment, he answered me in the negative; but that others assured him they did.

Young broods of swallows began to appear this year on July the eleventh, and young martins (*hirundines urbicæ*) were then fledged in their nests. Both species will breed again once. For I see by my *Fauna* of last year, that young broods come forth so late as September the eighteenth. Are not these late hatchings more in favour of hiding than migration? Nay, some young martins remained in their nests last year so late as September the twenty-ninth; and yet they totally disappeared with us by the fifth of October.

How strange is it that the swift, which seems to live exactly the same life with the swallow and house-martin, should leave us before the middle of August invariably! while the latter stay often till the middle of October; and once I saw numbers of house-martins on the seventh of November. The martins and red-wing fieldfares were flying in sight together; an uncommon assemblage of summer and winter birds.

A little yellow bird (it is either a species of the *alauda trivialis*, or rather perhaps of the *motacilla trochilus*[22]) still continues to make a sibilous shivering noise in the tops of tall woods. The *stoparola* of Ray (for which we have as yet no name in these parts) is called, in your *Zoology*, the fly-catcher.[23] There is one circumstance characteristic of this bird, which seems to have escaped observation, and that is, that it takes its stand on the top of some stake or post, from whence it springs forth on its prey, catching a fly in the air, and hardly ever touching the ground, but returning still to the same stand for many times together.

I perceive there are more than one species of the *motacilla trochilus*: Mr Derham supposes, in Ray's *Philos. Letters*, that he has discovered three. In these there is again an instance of some very common birds that have as yet no English name.

Mr Stillingfleet makes a question whether the black-cap (*motacilla atricapilla*) be a bird of passage or not: I think there is no doubt of it: for, in April, in the very first fine weather, they come trooping, all at once, into these parts, but are never seen in the winter. They are delicate songsters.[24]

Numbers of snipes breed every summer in some moory ground on the verge of this parish. It is very amusing to see the cock bird on wing at that time, and to hear his piping and humming notes.

I have had no opportunity yet of procuring any of those mice which I mentioned to you in town. The person that brought me the last says they are plenty in harvest, at which time I will take care to get more; and will endeavour to put the matter out of doubt, whether it be a nondescript species or not.

I suspect much there may be two species of water-rats. Ray says, and Linnæus after him, that the water-rat is web-footed behind. Now I have discovered a rat on the banks of our little stream that is not web-footed, and yet is an excellent swimmer and diver:[25] it answers exactly to the *mus amphibius* of Linnæus (see *Syst. Nat.*), which he says '*natat in fossis et urinatur.*' I should be glad to procure one '*plantis palmatis.*' Linnæus seems to be in a puzzle about his *mus amphibius*, and to doubt whether it differs from his *mus terrestris*; which if it be, as he allows, the '*mus agrestis capite grandi brachyuros*' of Ray, is widely different from the water-rat, both in size, make, and manner of life.

As to the *falco*,[26] which I mentioned in town, I shall take the liberty to send it down to you into Wales; presuming on your candour, that you will excuse me if it should appear as familiar to you as it is strange to me. Though mutilated '*qualem dices . . . antehac fuisse, tales cum sint reliquiæ!*'

It haunted a marshy piece of ground in quest of wild-ducks and snipes: but, when it was shot, had just knocked down a rook, which it was tearing in pieces. I cannot make it answer to any of our English hawks; neither could I find any like it at the curious exhibition of stuffed birds in Spring-gardens. I found it nailed up at the end of a barn, which is the countryman's museum.

The parish I live in is a very abrupt, uneven country, full of hills and woods, and therefore full of birds.

LETTER XI

Selborne, September 9, 1767.

IT will not be without impatience, that I shall wait for your thoughts with regard to the *falco*; as to its weight, breadth, etc., I wish I had set them down at the time; but, to the best of my remembrance, it weighed two pounds and eight ounces, and measured, from wing to wing, thirty-eight inches. Its cere and feet were yellow, and the circle of its eyelids bright yellow. As it had been killed some days, and the eyes were sunk, I could make no good observation on the colour of the pupils and the irides.

The most unusual birds I ever observed in these parts were a pair of hoopoes (*upupa*) which came several years ago in the summer, and frequented an ornamented piece of ground, which joins to my garden, for some weeks. They used to march about in a stately manner, feeding in the walks, many times in the day; and seemed disposed to breed in my outlet; but were frightened and persecuted by idle boys, who would never let them be at rest.

Three gross-beaks[27] (*loxia coccothraustes*) appeared some years ago in my fields, in the winter; one of which I shot: since that, now and then one is occasionally seen in the same dead season.

A cross-bill[28] (*loxia curvirostra*) was killed last year in this neighbourhood.

Our streams, which are small, and rise only at the end of the village, yield nothing but the bull's head or miller's thumb (*gobius fluviatilis capitatus*), the trout (*trutta fluviatilis*), the eel (*anguilla*), the lampern (*lampætra parva et fluviatilis*), and the stickle-back (*pisciculus aculeatus*).

We are twenty miles from the sea, and almost as many from a great river, and therefore see but little of sea-birds. As to wild fowls, we have a few teams of ducks bred in the moors where the snipes breed; and multitudes of widgeons and teals in hard weather frequent our lakes in the forest.

Having some acquaintance with a tame brown owl, I find that it casts up the fur of mice, and the feathers of birds in pellets, after the manner of hawks: when full, like a dog, it hides what it cannot eat.

The young of the barn-owl are not easily raised, as they want a constant supply of fresh mice: whereas the young of the brown owl will eat indiscriminately all that is brought; snails, rats, kittens, puppies, magpies, and any kind of carrion or offal.

The house-martins have eggs still, and squab-young. The last swift I observed was about the twenty-first of August; it was a straggler.

Red-starts, fly-catchers, white-throats, and *reguli non cristati*, still appear; but I have seen no black-caps lately.

I forgot to mention that I once saw, in Christ Church College quadrangle in Oxford, on a very sunny warm morning, a house-martin flying about, and settling on the parapet, so late as the twentieth of November.

At present I know only two species of bats,[29] the common *vespertilio murinus* and the *vespertilio auritus*.

I was much entertained last summer with a tame bat, which would take flies out of a person's hand. If you gave it anything to eat, it brought its wings round before the mouth,[30] hovering and hiding its head in the manner of birds of prey when they feed. The adroitness it showed in shearing off the wings of the flies, which were always rejected, was worthy of observation, and pleased me much. Insects seem to be most acceptable, though it did not refuse raw flesh when offered: so that the notion that bats go down chimnies and gnaw men's bacon, seems no improbable story. While I amused myself with this wonderful quadruped, I saw it several times confute the vulgar opinion, that bats when down on a flat surface cannot get on the wing again, by rising with great ease from the floor. It ran,

I observed, with more dispatch than I was aware of; but in a most ridiculous and grotesque manner.

Bats drink on the wing, like swallows, by sipping the surface, as they play over pools and streams. They love to frequent waters, not only for the sake of drinking, but on account of insects, which are found over them in the greatest plenty. As I was going, some years ago, pretty late, in a boat from Richmond to Sunbury, on a warm summer's evening, I think I saw myriads of bats between the two places: the air swarmed with them all along the Thames, so that hundreds were in sight at a time.

I am, etc.

LETTER XII

November 4, 1767.

SIR,

It gave me no small satisfaction to hear that the *falco*★ turned out an uncommon one. I must confess I should have been better pleased to have heard that I had sent you a bird that you had never seen before; but that, I find, would be a difficult task.

I have procured some of the mice mentioned in my former letters, a young one and a female with young, both of which I have preserved in brandy. From the colour, shape, size, and manner of nesting, I make no doubt but that the species is nondescript.[31] They are much smaller and more slender than the *mus domesticus medius* of Ray; and have more of the squirrel or dormouse colour: their belly is white, a straight line along their sides divides the shades of their back and belly. They never enter into houses; are carried into ricks and barns with the sheaves; abound in harvest, and build their nests amidst the straws of the corn above the ground, and sometimes in thistles.

★ This hawk proved to be the *falco peregrinus*; a variety.

They breed as many as eight at a litter, in a little round nest composed of the blades of grass or wheat.

One of these nests I procured this autumn, most artificially platted, and composed of the blades of wheat; perfectly round, and about the size of a cricket-ball; with the aperture so ingeniously closed, that there was no discovering to what part it belonged. It was so compact and well filled, that it would roll across the table without being discomposed, though it contained eight little mice that were naked and blind. As this nest was perfectly full, how could the dam come at her litter respectively so as to administer a teat to each? perhaps she opens different places for that purpose, adjusting them again when the business is over: but she could not possibly be contained herself in the ball with her young, which moreover would be daily increasing in bulk. This wonderful procreant cradle, an elegant instance of the efforts of instinct, was found in a wheat-field, suspended in the head of a thistle.

A gentleman, curious in birds, wrote me word that his servant had shot one last January, in that severe weather, which he believed would puzzle me. I called to see it this summer, not knowing what to expect: but, the moment I took it in hand, I pronounced it the male *garrulus bohemicus* or German silk-tail,[32] from the five peculiar crimson tags or points which it carries at the end of five of the short remiges. It cannot, I suppose, with any propriety, be called an English bird: and yet I see, by Ray's *Philosoph. Letters*, that great flocks of them, feeding upon haws, appeared in this kingdom in the winter of 1685.

The mention of haws put me in mind that there is a total failure of that wild fruit, so conducive to the support of many of the winged nation. For the same severe weather, late in the spring, which cut off all the produce of the more tender and curious trees, destroyed also that of the more hardy and common.

Some birds, haunting with the missel-thrushes, and feeding

on the berries of the yew-tree, which answered to the description of the *merula torquata*, or ring-ousel, were lately seen in this neighbourhood. I employed some people to procure me a specimen, but without success. See Letter XX.

Query. – Might not canary birds be naturalized to this climate, provided their eggs were put in the spring, into the nests of some of their congeners, as goldfinches, greenfinches, etc.? Before winter perhaps they might be hardened, and able to shift for themselves.

About ten years ago I used to spend some weeks yearly at Sunbury, which is one of those pleasant villages lying on the Thames, near Hampton-court. In the autumn, I could not help being much amused with those myriads of the swallow kind which assemble in those parts. But what struck me most was, that, from the time they began to congregate, forsaking the chimnies and houses, they roosted every night in the osier-beds of the aits of that river. Now this resorting towards that element, at that season of the year, seems to give some countenance to the northern opinion (strange as it is) of their retiring under water. A Swedish naturalist is so much persuaded of that fact, that he talks, in his calendar of *Flora*,[33] as familiarly of the swallows going under water in the beginning of September, as he would of his poultry going to roost a little before sunset.

An observing gentleman in London writes me word that he saw a house-martin, on the twenty-third of last October, flying in and out of its nest in the Borough. And I myself, on the twenty-ninth of last October (as I was travelling through Oxford), saw four or five swallows hovering round and settling on the roof of the county-hospital.

Now is it likely that these poor little birds (which perhaps had not been hatched but a few weeks) should, at that late season of the year, and from so midland a county, attempt a voyage to Goree or Senegal, almost as far as the equator?*

I acquiesce entirely in your opinion – that, though most of

*See Adamson's *Voyage to Senegal*.

38

the swallow kind may migrate, yet that some do stay behind and hide with us during the winter.

As to the short-winged soft-billed birds, which come trooping in such numbers in the spring, I am at a loss even what to suspect about them. I watched them narrowly this year, and saw them abound till about Michaelmas, when they appeared no longer. Subsist they cannot openly among us, and yet elude the eyes of the inquisitive: and, as to their hiding, no man pretends to have found any of them in a torpid state in the winter. But with regard to their migration, what difficulties attend that supposition! that such feeble bad fliers (who the summer long never flit but from hedge to hedge) should be able to traverse vast seas and continents in order to enjoy milder seasons amidst the regions of Africa!

LETTER XIII

Selborne, Jan. 22, 1768.

SIR,

As in one of your former letters you expressed the more satisfaction from my correspondence on account of my living in the most southerly county; so now I may return the compliment, and expect to have my curiosity gratified by your living much more to the north.

For many years past I have observed that towards Christmas vast flocks of chaffinches have appeared in the fields; many more, I used to think, than could be hatched in any one neighbourhood. But, when I came to observe them more narrowly, I was amazed to find that they seemed to be almost all hens. I communicated my suspicions to some intelligent neighbours, who, after taking pains about the matter, declared that they also thought them all mostly females; at least fifty to one. This extraordinary occurrence brought to my mind the remark of

Linnæus; that 'before winter, all their hen chaffinches migrate through Holland into Italy.' Now I want to know, from some curious person in the north, whether there are any large flocks of these finches with them in the winter, and of which sex they mostly consist? For, from such intelligence, one might be able to judge whether our female flocks migrate from the other end of the island, or whether they come over to us from the continent.

We have, in the winter, vast flocks of the common linnets; more, I think, than can be bred in any one district. These, I observe, when the spring advances, assemble on some tree in the sunshine, and join all in a gentle sort of chirping, as if they were about to break up their winter quarters and betake themselves to their proper summer homes. It is well known, at least, that the swallows and the fieldfares do congregate with a gentle twittering before they make their respective departure.

You may depend on it that the bunting, *emberiza miliaria*,[34] does not leave this country in the winter. In January 1767 I saw several dozen of them, in the midst of a severe frost, among the bushes on the downs near Andover: in our woodland enclosed district it is a rare bird.

Wagtails, both white and yellow,[35] are with us all the winter. Quails crowd to our southern coast, and are often killed in numbers by people that go on purpose.

Mr Stillingfleet, in his Tracts, says that 'if the wheatear (*œanthe*) does not quit England, it certainly shifts places; for about harvest they are not to be found, where there was before great plenty of them.' This well accounts for the vast quantities that are caught about that time on the south downs near Lewes, where they are esteemed a delicacy. There have been shepherds, I have been credibly informed, that have made many pounds in a season by catching them in traps. And though such multitudes are taken, I never saw (and I am well acquainted with those parts) above two or three at a time: for they are never gregarious. They may, perhaps, migrate in general; and, for that purpose, draw towards the coast of Sussex in autumn; but that they do not all withdraw I am sure; because I see a few stragglers in many counties, at all times of the year, especially about warrens and stone quarries.

I have no acquaintance, at present, among the gentlemen of the navy: but have written to a friend, who was a sea-chaplain in the late war, desiring him to look into his minutes, with respect to birds that settled on their rigging during their voyage up or down the channel. What Hasselquist[36] says on that subject is remarkable: there were little short-winged birds frequently coming on board his ship all the way from our channel quite up to the Levant, especially before squally weather.

What you suggest, with regard to Spain, is highly probable. The winters of Andalusia are so mild, that, in all likelihood, the soft-billed birds that leave us at that season may find insects sufficient to support them there.

Some young man, possessed of fortune, health, and leisure, should make an autumnal voyage into that kingdom; and should spend a year there, investigating the natural history of that vast country. Mr Willughby* passed through that kingdom on such an errand; but he seems to have skirted along in a superficial manner and an ill humour, being much disgusted at the rude, dissolute manners of the people.

I have no friend left now at Sunbury to apply to about the swallows roosting on the aits of the Thames: nor can I hear any more about those birds which I suspected were *merulæ torquatæ*.

As to the small mice, I have farther to remark, that though they hang their nests for breeding up amidst the straws of the standing corn, above the ground; yet I find that, in the winter, they burrow deep in the earth, and make warm beds of grass: but their grand rendezvous seems to be in corn-ricks, into which they are carried at harvest. A neighbour housed an oat-rick lately, under the thatch of which were assembled near an hundred, most of which were taken; and some I saw. I meas-

*See Ray's *Travels*, p. 466.

ured them; and found that, from nose to tail, they were just two inches and a quarter, and their tails just two inches long. Two of them, in a scale, weighed down just one copper halfpenny, which is about a third of an ounce avoirdupois: so that I suppose they are the smallest quadrupeds in this island.[37] A full-grown *mus medius domesticus* weighs, I find, one ounce, lumping weight, which is more than six times as much as the mouse above; and measures from nose to rump four inches and a quarter, and the same in its tail.

We have had a very severe frost and deep snow this month. My thermometer was one day fourteen degrees and a half below the freezing point, within doors. The tender evergreens were injured pretty much. It was very providential that the air was still, and the ground well covered with snow, else vegetation in general must have suffered prodigiously. There is reason to believe that some days were more severe than any since the year 1739–40.

I am, etc., etc.

LETTER XIV

Selborne, March 12, 1768.

DEAR SIR,

If some curious gentleman would procure the head of a fallow-deer, and have it dissected, he would find it furnished with two *spiracula*, or breathing-places,[38] beside the nostrils; probably analogous to the *puncta lachrymalia* in the human head. When the deer are thirsty they plunge their noses, like some horses, very deep under water, while in the act of drinking, and continue them in that situation for a considerable time, but, to obviate any inconveniency, they can open two vents, one at the inner corner of each eye, having a communication with the nose. Here seems to be an extraordinary provision of nature worthy our attention; and which has not, that I know of, been noticed by any naturalist. For it looks as if these creatures would not be suffocated, though both their mouths and nostrils were stopped. This curious formation of the head may be of singular service to beasts of chase, by affording them free respiration: and no doubt these additional nostrils are thrown open when they are hard run.* Mr Ray observed that, at Malta, the owners slit up the nostrils of such asses as were hard worked: for they, being naturally strait or small, did not admit air sufficient to serve them when they travelled or laboured in that hot climate. And we know that grooms, and gentlemen of the turf, think large nostrils necessary, and a perfection, in hunters and running horses.

Oppian, the Greek poet, by the following line, seems to have had some notion that stags have four *spiracula*:

Τετράδυμοι ῥῖνες, πίσυρες πνοιῇσι δίαυλοι
Quadrifidæ nares, quadruplices ad respirationem canales.
                                    OPP. *Cyn.* lib. ii. 1. 181.

Writers, copying from one another, make Aristotle say that goats breathe at their ears; whereas he asserts just the contrary: '᾽Αλκμαιὼν γὰρ οὐκ ἀληθῆ λέγει, φάμενος ἀναπνεῖν τὰς αἶγας κατὰ τὰ ὦτα.' 'Alcmæon does not advance what is true, when he avers that goats breathe through their ears.' – *History of Animals*. Book I. chap. xi.

---

*In answer to this account, Mr Pennant sent me the following curious and pertinent reply: 'I was much surprised to find in the antelope something analogous to what you mention as so remarkable in deer. This animal has a long slit beneath each eye, which can be opened and shut at pleasure. On holding an orange to one, the creature made as much use of those orifices as of his nostrils, applying them to the fruit, and seeming to smell it through them.'

LETTER XV

Selborne, March 30, 1768

DEAR SIR,

Some intelligent country people have a notion that we have, in these parts, a species of the *genus mustelinum*, besides the weasel, stoat, ferret, and polecat; a little reddish beast, not much bigger than a field mouse, but much longer, which they call a *cane*.[39] This piece of intelligence can be little depended on; but farther inquiry may be made.

A gentleman in this neighbourhood had two milk-white rooks in one nest. A booby of a carter, finding them before they were able to fly, threw them down and destroyed them, to the regret of the owner, who would have been glad to have preserved such a curiosity in his rookery. I saw the birds myself nailed against the end of a barn, and was surprised to find that their bills, legs, feet, and claws were milk-white.

A shepherd saw, as he thought, some white larks on a down above my house this winter: were not these the *emberiza nivalis*, the snow-flake[40] of the *Brit. Zool.*? No doubt they were.

A few years ago I saw a cock bullfinch in a cage, which had been caught in the fields after it had come to its full colours. In about a year it began to look dingy; and, blackening every succeeding year, it became coal-black at the end of four. Its chief food was hemp-seed. Such influence has food on the colour of animals! The pied and mottled colours of domesticated animals are supposed to be owing to high, various, and unusual food.

I had remarked, for years, that the root of the cuckoo-pint (*arum*) was frequently scratched out of the dry banks of hedges, and eaten in severe snowy weather. After observing, with some exactness, myself, and getting others to do the same, we found it was the thrush kind that searched it out. The root of the *arum* is remarkably warm and pungent.

Our flocks of female chaffinches have not yet forsaken us.

The blackbirds and thrushes are very much thinned down by that fierce weather in January.

In the middle of February I discovered, in my tall hedges, a little bird that raised my curiosity: it was of that yellow-green colour that belongs to the *salicaria* kind, and, I think, was soft-billed. It was no *parus*; and was too long and too big for the golden-crowned wren, appearing most like the largest willow-wren. It hung sometimes with its back downwards, but never continuing one moment in the same place.[41] I shot at it, but it was so desultory that I missed my aim.

I wonder that the stone curlew, *charadrius œdicnemus*, should be mentioned by the writers as a rare bird: it abounds in all the campaign parts of Hampshire and Sussex, and breeds, I think, all the summer, having young ones, I know, very late in the autumn. Already they begin clamouring in the evening. They cannot, I think, with any propriety, be called, as they are by Mr Ray, '*circa aquas versantes*'; for with us, by day at least, they haunt only the most dry, open, upland fields and sheep walks, far removed from water. What they may do in the night I cannot say. Worms are their usual food, but they also eat toads and frogs.

I can show you some good specimens of my new mice. Linnæus, perhaps, would call the species *mus minimus*.

LETTER XVI

Selborne, April 18, 1768.

DEAR SIR,

The history of the stone curlew, *charadrius œdicnemus* is as follows. It lays its eggs, usually two, never more than three, on the bare ground, without any nest, in the field; so that the coun-tryman, in stirring his fallows, often destroys them. The young run immediately from the egg like partridges, etc., and are withdrawn to some flinty field by their dam, where they skulk among the stones, which are their best security; for their feathers are so exactly of the colour of our grey spotted flints, that the most exact observer, unless he catches the eye of the young bird, may be eluded. The eggs are short and round; of a dirty white, spotted with dark bloody blotches. Though I might not be able, just when I pleased, to procure you a bird, yet I could show you them almost any day; and any evening you may hear them round the village, for they make a clamour which may be heard a mile. Œdicnemus is a most apt and expressive name for them, since their legs seem swoln like those of a gouty man. After harvest I have shot them before the pointers in turnip-fields.

I make no doubt but there are three species of the willow-wrens:[42] two I know perfectly; but have not been able yet to procure the third. No two birds can differ more in their notes, and that constantly, than those two that I am acquainted with; for the one has a joyous, easy, laughing note; the other a harsh loud chirp. The former is every way larger, and three-quarters of an inch longer, and weighs two drams and a half; while the latter weighs but two: so the songster is one-fifth heavier than the chirper. The chirper (being the first summer-bird of passage that is heard, the wryneck sometimes excepted) begins his two notes in the middle of March, and continues them through the spring and summer till the end of August, as appears by my journals. The legs of the larger of these two are flesh-coloured; of the less, black.

The grasshopper-lark[43] began his sibilous note in my fields last Saturday. Nothing can be more amusing than the whisper of this little bird, which seems to be close by though at an hundred yards distance; and, when close at your ear, is scarce any louder than when a great way off. Had I not been a little acquainted with insects, and known that the grasshopper kind is not yet hatched, I should have hardly believed but that it had been a *locusta* whispering in the bushes. The country people laugh when you tell them that it is the note of a bird. It is a most artful creature, skulking in the thickest part of a bush; and will sing at a yard distance, provided it be concealed. I was obliged to get a person to go on the other side of the hedge where it haunted; and then it would run, creeping like a mouse, before us for a hundred yards together, through the bottom of the thorns; yet it would not come into fair sight: but in a morning early, and when undisturbed, it sings on the top of a twig, gaping and shivering with its wings. Mr Ray himself had no knowledge of this bird, but received his account from Mr Johnson, who apparently confounds it with the *reguli non cristati*, from which it is very distinct. See Ray's *Philosophical Letters*, p. 108.

A LIST *of the* SUMMER BIRDS *of* PASSAGE *discovered in this neighbourhood, ranged somewhat in the Order in which they appear:*

	Linnæi Nomina.
Smallest willow-wren,	*Motacilla trochilus.*
Wryneck,	*Jynx torquilla.*
House-swallow,	*Hirundo rustica.*
Martin,	*Hirundo urbica.*
Sand-martin,	*Hirundo riparia.*
Cuckoo,	*Cuculus canorus.*
Nightingale,	*Motacilla luscinia.*
Black-cap,	*Motacilla atricapilla.*
White-throat,	*Motacilla sylvia.*
Middle willow-wren,	*Motacilla trochilus.*
Swift,	*Hirundo apus.*
Stone curlew?	*Charadrius œdicnemus?*
Turtle-dove?	*Turtur aldrovandi?*
Grasshopper-lark,	*Alauda trivialis.*
Landrail,	*Rallus crex.*
Largest willow-wren,	*Motacilla trochilus.*
Redstart,	*Motacilla phœnicurus.*
Goat-sucker, or fern-owl,	*Caprimulgus europæus.*
Fly-catcher,	*Muscicapa grisola.*

The fly-catcher (*stoparola*) has not yet appeared: it usually breeds in my vine. The redstart begins to sing: its note is short and imperfect, but is continued till about the middle of June. The willow-wrens (the smaller sort) are horrid pests in a garden, destroying the pease, cherries, currants, etc., and are so tame that a gun will not scare them.

My countrymen talk much of a bird that makes a clatter with its bill against a dead bough, or some old pales, calling it a jar-bird. I procured one to be shot in the very fact; it proved to be the *sitta europæa* (the nut-hatch). Mr Ray says that the less spotted woodpecker does the same. This noise may be heard a furlong or more.

Now is the only time to ascertain the short-winged summer birds; for, when the leaf is out, there is no making any remarks on such a restless tribe; and, when once the young begin to appear, it is all confusion: there is no distinction of genus, species, or sex.

In breeding-time snipes play over the moors, piping and humming: they always hum as they are descending. Is not their hum ventriloquous like that of a turkey? Some suspect it is made by their wings.[44]

This morning I saw the golden-crowned wren,[45] whose crown glitters like burnished gold. It often hangs like a tit-mouse, with its back downwards.

Yours, etc., etc.

LETTER XVII

Selborne, June 18, 1768.

DEAR SIR,

On Wednesday last arrived your agreeable letter of June the 10th. It gives me great satisfaction to find that you pursue these studies still with such vigour, and are in such forwardness with regard to reptiles and fishes.

The reptiles, few as they are, I am not acquainted with, so well as I could wish, with regard to their natural history. There is a degree of dubiousness and obscurity attending the propagation of this class of animals, sometimes analogous to that of the *cryptogamia* in the sexual system of plants: and the case is the same as regards some of the fishes: as the eel, etc.

The method in which toads procreate and bring forth seems to me very much in the dark. Some authors say that they are viviparous: and yet Ray classes them among his oviparous animals; and is silent with regard to the manner of their bringing forth. Perhaps they may be ἔσω μὲν ᾠοτόκοι, ἔξω δὲ ζωοτόκοι, as is known to be the case with the viper.

The copulation of frogs (or at least the appearance of it; for Swammerdam[46] proves that the male has no *penis intrans*) is notorious to everybody: because we see them sticking upon each other's backs for a month together in spring: and yet I never saw, or read, of toads being observed in the same situation.[47] It is strange that the matter with regard to the venom of toads has not yet been settled. That they are not noxious to some animals is plain: for ducks, buzzards, owls, stone curlews, and snakes, eat them, to my knowledge, with impunity. And I well remember the time, but was not eye-witness to the fact (though numbers of persons were), when a quack, at this village, ate a toad to make the country people stare; afterwards he drank oil.

I have been informed also, from undoubted authority, that some ladies (ladies you will say of peculiar taste) took a fancy to a toad, which they nourished summer after summer, for many years, till he grew to a monstrous size, with the maggots which turn to flesh flies. The reptile used to come forth every evening from an hole under the garden-steps; and was taken up, after supper, on the table to be fed. But at last a tame raven, kenning him as he put forth his head, gave him such a severe stroke with his horny beak as put out one eye. After this accident the creature languished for some time and died.

I need not remind a gentleman of your extensive reading of the excellent account there is from Mr Derham, in Ray's *Wisdom of God in the Creation* (p. 365), concerning the migration of frogs from their breeding ponds. In this account he at once subverts that foolish opinion of their dropping from the clouds in rain; showing that it is from the grateful coolness and moisture of those showers that they are tempted to set out on their travels, which they defer till those fall. Frogs are as yet in their tadpole state; but in a few weeks, our lanes, paths, fields, will swarm for a few days with myriads of these emigrants, no larger than my little finger nail. Swammerdam gives a most accurate account of the method and situation in which the male impregnates the spawn of the female. How wonderful is the œconomy of Providence with regard to the limbs of so vile a reptile! While it is aquatic it has a fish-like tail, and no legs: as soon as the legs sprout, the tail drops off as useless,[48] and the animal betakes itself to the land.

Merret, I trust, is widely mistaken when he advances that the *rana arborea*[49] is an English reptile; it abounds in Germany and Switzerland.

It is to be remembered that the *salamandra aquatica* of Ray (the water-newt or eft) will frequently bite at the angler's bait, and is often caught on his hook. I used to take it for granted that the *salamandra aquatica* was hatched, lived, and died in the water. But John Ellis, Esq., F.R.S. (the coralline Ellis),[50] asserts, in a letter to the Royal Society, dated June 5th, 1766, in his account of the *mud inguana*, an amphibious *bipes*, from South Carolina, that the water-eft, or newt,[51] is only the larva of the land-eft, as tadpoles are of frogs. Lest I should be suspected to misunderstand his meaning, I shall give it in his own words. Speaking of the *opercula* or covering to the gills of the *mud inguana*, he proceeds to say that 'The forms of these pennated coverings approach very near to what I have some time ago observed in the larva or aquatic state of our English *lacerta*, known by the name of eft, or newt; which serve them for coverings to their gills, and for fins to swim with while in this state; and which they lose, as well as the fins of their tails, when they change their state, and become land animals, as I have observed, by keeping them alive for some time myself.'

Linnæus, in his *Systema Naturæ*, hints at what Mr Ellis advances more than once.

Providence has been so indulgent to us as to allow of but one venomous reptile of the serpent kind in these kingdoms, and that is the viper. As you propose the good of mankind to be an object of your publications, you will not omit to mention common salad-oil as a sovereign remedy against the bite of the viper. As to the blind worm (*anguis fragilis*, so called because it snaps in sunder with a small blow), I have found, on examination, that it is perfectly innocuous. A neighbouring yeoman (to whom I am indebted for some good hints) killed and opened a female viper about the twenty-seventh of May: he found her filled with a chain of eleven eggs, about the size of those of a blackbird; but none of them were advanced so far towards a state of maturity as to contain any rudiments of young. Though they are oviparous, yet they are viviparous also, hatching their young within their bellies, and then bringing them forth. Whereas snakes lay chains of eggs[52] every summer in my melon beds, in spite of all that my people can do to prevent them; which eggs do not hatch till the spring following, as I have often experienced. Several intelligent folks assure me that they have seen the viper open her mouth and admit her helpless young down her throat on sudden surprises, just as the female opossum does her brood into the pouch under her belly, upon the like emergencies; and yet the London viper-catchers insist on it, to Mr Barrington, that no such thing ever happens. The serpent kind eat, I believe, but once in a year; or rather, but only just at one season of the year. Country people talk much of a water-snake, but I am pretty sure, without any reason; for the common snake (*coluber natrix*) delights much to sport in the water, perhaps with a view to procure frogs and other food.

I cannot well guess how you are to make out your twelve species of reptiles, unless it be by the various species, or rather varieties, of our *lacerti*,[53] of which Ray enumerates five. I have not had an opportunity of ascertaining these; but remember well to have seen, formerly, several beautiful green *lacerti* on the sunny sandbanks near Farnham, in Surrey; and Ray admits there are such in Ireland.

LETTER XVIII

Selborne, July 27, 1768.

DEAR SIR,

I received your obliging and communicative letter of June the 28th, while I was on a visit at a gentleman's house, where I had neither books to turn to, nor leisure to sit down, to return you an answer to many queries, which I wanted to resolve in the best manner that I am able.

A person, by my order, has searched our brooks, but could find no such fish as the *gasterosteus pungitius*:[54] he found the *gasterosteus aculeatus* in plenty. This morning, in a basket, I packed a little earthen pot full of wet moss, and in it some sticklebacks, male and female; the females big with spawn: some lamperns; some bull's heads; but I could produce no minnows. This basket will be in Fleet-street by eight this evening; so I hope Mazel will have them fresh and fair to-morrow morning. I gave some directions, in a letter, to what particulars the engraver should be attentive.

Finding, while I was on a visit, that I was within a reasonable distance of Ambresbury, I sent a servant over to that town, and procured several living specimens of loaches, which he brought, safe and brisk, in a glass decanter. They were taken in the gullies that were cut for watering the meadows. From these fishes (which measured from two to four inches in length) I took the following description: 'The loach, in its general aspect, has a pellucid appearance: its back is mottled with irregular collections of small black dots, not reaching much below the *linea lateralis*, as are the back and tail fins: a black line runs from each eye down to the nose; its belly is of a silvery white; the upper jaw projects beyond the lower, and is surrounded with six feelers, three on each side; its pectoral fins are large, its ventral much smaller; the fin behind its anus small; its dorsal fin large, containing eight spines; its tail, where it joins to the tail-fin, remarkably broad, without any taperness, so as to be characteristic of this genus: the tail-fin is broad, and square at the end. From the breadth and muscular strength of the tail, it appears to be an active nimble fish.'

In my visit I was not very far from Hungerford, and did not forget to make some inquiries concerning the wonderful method of curing cancers by means of toads. Several intelligent persons, both gentry and clergy, do, I find, give a great deal of credit to what was asserted in the papers: and I myself dined with a clergyman who seemed to be persuaded that what is related is matter of fact; but, when I came to attend to his account, I thought I discerned circumstances which did not a little invalidate the woman's story of the manner in which she came by her skill. She says of herself 'that, labouring under a virulent cancer, she went to some church where there was a vast crowd: on going into a pew, she was accosted by a strange clergyman; who, after expressing compassion for her situation, told her that if she would make such an application of living toads as is mentioned she would be well.' Now is it likely that this unknown gentleman should express so much tenderness for this single sufferer, and not feel any for the many thousands that daily languish under this terrible disorder? Would he not have made use of this invaluable nostrum for his own emolument; or, at least, by some means of publication or other, have found a method of making it public for the good of mankind? In short, this woman (as it appears to me) having set up for a cancer-doctress, finds it expedient to amuse the country with this dark and mysterious relation.

The water-eft has not, that I can discern, the least appearance of any gills; for want of which it is continually rising to the surface of the water to take in fresh air. I opened a big-bellied one indeed, and found it full of spawn. Not that this circumstance at all invalidates the assertion that they are larvæ: for the larvæ of insects are full of eggs, which they exclude the instant they enter their last state. The water-eft is continually climbing over the brims of the vessel, within which we keep it in water, and wandering away: and people every summer see numbers crawling out of the pools where they are hatched, up the dry banks. There are varieties of them, differing in colour; and some have fins up their tail and back, and some have not.

LETTER XIX

Selborne, Aug. 17, 1768.

DEAR SIR,

I have now, past dispute, made out three distinct species of the willow-wrens (*motacillæ trochili*) which constantly and invariably use distinct notes. But, at the same time, I am obliged to confess that I know nothing of your willow-lark.* In my letter of April the 18th, I told you peremptorily that I knew your willow-lark, but had not seen it then: but, when I came to procure it, it proved, in all respects, a very *motacilla trochilus*; only that it is a size larger than the two other, and the yellow-green of the whole upper part of the body is more vivid, and the belly of a clearer white. I have specimens of the three sorts now lying before me; and can discern that there are three gradations of sizes, and that the least has black legs, and the other two flesh-coloured ones. The yellowest bird is consider-

Brit. Zool. edit. 1776, octavo, p. 381.

ably the largest, and has its quill-feathers and secondary feathers tipped with white, which the others have not. This last haunts only the tops of trees in high beechen woods, and makes a sibilous grasshopper-like noise, now and then, at short intervals, shivering a little with its wings when it sings; and is, I make no doubt now, the *regulus non cristatus* of Ray, which he says '*cantat voce stridulâ locustæ.*' Yet this great ornithologist never suspected that there were three species.

LETTER XX

Selborne, October 8, 1768.

IT is, I find, in zoology as it is in botany: all nature is so full, that that district produces the greatest variety which is the most examined. Several birds, which are said to belong to the north only, are, it seems, often in the south. I have discovered this summer three species of birds with us, which writers mention as only to be seen in the northern counties. The first that was brought me (on the 14th of May) was the sandpiper, *tringa hypoleucus*: it was a cock bird, and haunted the banks of some ponds near the village; and, as it had a companion, doubtless intended to have bred near that water. Besides, the owner has told me since, that, on recollection, he has seen some of the same birds round his ponds in former summers.

The next bird that I procured (on the 21st of May) was a male red-backed butcher bird,[55] *lanius collurio*. My neighbour, who shot it, says that it might easily have escaped his notice, had not the outcries and chattering of the white-throats and other small birds drawn his attention to the bush where it was: its craw was filled with the legs and wings of beetles.

The next rare birds (which were procured for me last week) were some ring-ousels, *turdi jorquati*.

This week twelve months a gentleman from London, being

with us, was amusing himself with a gun, and found, he told us, on an old yew hedge where there were berries, some birds like blackbirds, with rings of white round their necks: a neighbouring farmer also at the same time observed the same; but, as no specimens were procured little notice was taken. I mentioned this circumstance to you in my letter of November the 4th, 1767 (you, however, paid but small regard to what I said, as I had not seen these birds myself); but last week, the aforesaid farmer, seeing a large flock, twenty or thirty of these birds, shot two cocks and two hens: and says, on recollection, that he remembers to have observed these birds again last spring, about Lady-day, as it were, on their return to the north. Now perhaps these ousels are not the ousels of the north of England, but belong to the more northern parts of Europe; and may retire before the excessive rigour of the frosts in those parts; and return to breed in the spring, when the cold abates. If this be the case, here is discovered a new bird of winter passage, concerning whose migrations the writers are silent: but if these birds should prove the ousels of the north of England, then here is a migration disclosed within our own kingdom never before remarked. It does not yet appear whether they retire beyond the bounds of our island to the south; but it is most probable that they usually do, or else one cannot suppose that they would have continued so long unnoticed in the southern counties. The ousel is larger than a blackbird, and feeds on haws; but last autumn (when there were no haws) it fed on yew-berries: in the spring it feeds on ivy-berries, which ripen only at that season, in March and April.

I must not omit to tell you (as you have been so lately on the study of reptiles) that my people, every now and then of late, draw up with a bucket of water from my well, which is 63 feet deep, a large black warty lizard[56] with a fin-tail and yellow belly. How they first came down at that depth, and how they were ever to have got out thence without help, is more than I am able to say.

My thanks are due to you for your trouble and care in the examination of a buck's head. As far as your discoveries reach at present, they seem much to corroborate my suspicions; and I hope Mr——[57] may find reason to give his decision in my favour; and then, I think, we may advance this extraordinary provision of nature as a new instance of the wisdom of God in the creation.

As yet I have not quite done with my history of the *ædicnemus*, or stone curlew; for I shall desire a gentleman in Sussex (near whose house these birds congregate in vast flocks in the autumn) to observe nicely when they leave him (if they do leave him), and when they return again in the spring; I was with this gentleman lately, and saw several single birds.

LETTER XXI

Selborne, Nov. 28, 1768.

DEAR SIR,

With regard to the *ædicnemus*, or stone curlew, I intend to write very soon to my friend near Chichester, in whose neighbourhood these birds seem most to abound; and shall urge him to take particular notice when they begin to congregate, and afterwards to watch them most narrowly whether they do not withdraw themselves during the dead of the winter. When I have obtained information with respect to this circumstance, I shall have finished my history of the stone curlew; which I hope will prove to your satisfaction, as it will be, I trust, very near the truth. This gentleman, as he occupies a large farm of his own, and is abroad early and late, will be a very proper spy upon the motions of these birds: and besides, as I have prevailed on him to buy the *Naturalists' Journal*[58] (with which he is much delighted), I shall expect that he will be very exact in his dates.

It is very extraordinary, as you observe, that a bird so common with us should never straggle to you.[59]

And here will be the properest place to mention, while I think of it, an anecdote which the above-mentioned gentleman told me when I was last at his house; which was that, in a warren joining to his outlet, many daws (*corvi monedulæ*) build every year in the rabbit burrows under ground. The way he and his brothers used to take their nests, while they were boys, was by listening at the mouths of the holes; and, if they heard the young ones cry, they twisted the nest out with a forked stick. Some water-fowls (viz., the puffins) breed, I know, in that manner; but I should never have suspected the daws of building in holes on the flat ground.

Another very unlikely spot is made use of by daws as a place to breed in, and that is Stonehenge. These birds deposit their nests in the interstices between the upright and the impost stones of that amazing work of antiquity: which circumstance alone speaks the prodigious height of the upright stones, that they should be tall enough to secure those nests from the annoyance of shepherd-boys, who are always idling round that place.

One of my neighbours last Saturday, November the 26th, saw a martin in a sheltered bottom: the sun shone warm, and the bird was hawking briskly after flies. I am now perfectly satisfied that they do not all leave this island in the winter.

You judge very right, I think, in speaking with reserve and caution concerning the cures done by toads: for, let people advance what they will on such subjects, yet there is such a propensity in mankind towards deceiving and being deceived, that one cannot safely relate any thing from common report, especially in print, without expressing some degree of doubt and suspicion.

Your approbation, with regard to my new discovery of the migration of the ring-ousel, gives me satisfaction; and I find you concur with me in suspecting that they are foreign birds which visit us. You will be sure, I hope, not to omit to make inquiry whether your ring ousels leave your rocks in the autumn. What puzzles me most, is the very short stay they make with us; for in about three weeks they are all gone. I shall be very curious to remark whether they will call on us at their return in the spring, as they did last year.

I want to be better informed with regard to ichthyology. If fortune had settled me near the sea-side, or near some great river, my natural propensity would soon have urged me to have made myself acquainted with their productions: but as I have lived mostly in inland parts, and in an upland district, my knowledge of fishes extends little farther than to those common sorts which our brooks and lakes produce.

I am, etc.

LETTER XXII

Selborne, July 2, 1769.

DEAR SIR,

As to the peculiarity of jackdaws building with us under the ground in rabbit-burrows, you have, in part, hit upon the reason; for, in reality, there are hardly any towers or steeples in all this country. And perhaps, Norfolk excepted, Hampshire and Sussex are as meanly furnished with churches[60] as almost any counties in the kingdom. We have many livings of two or three hundred pounds a year, whose houses of worship make little better appearance than dovecots. When I first saw Northamptonshire, Cambridgeshire and Huntingdonshire, and the fens of Lincolnshire, I was amazed at the number of spires which presented themselves in every point of view. As an admirer of prospects, I have reason to lament this want in my own country; for such objects are very necessary ingredients in an elegant landscape.

What you mention with respect to reclaimed toads raises my curiosity. An ancient author, though no naturalist, has well remarked that 'Every kind of beasts, and of birds, and of serpents, and of things in the sea, is tamed, and hath been tamed, of mankind.'*

It is a satisfaction to me to find that a green lizard has actually been procured for you in Devonshire; because it corroborates my discovery, which I made many years ago, of the same sort, on a sunny sandbank near Farnham in Surrey. I am well acquainted with the south hams of Devonshire; and can suppose that district, from its southerly situation, to be a proper habitation for such animals in their best colours.

Since the ring-ousels of your vast mountains do certainly not forsake them against winter,[61] our suspicions that those which visit this neighbourhood about Michaelmas are not English birds, but driven from the more northern parts of Europe by the frosts, are still more reasonable: and it will be worth your pains to endeavour to trace from whence they come, and to inquire why they make so very short a stay.

In your account of your error with regard to the two species of herons, you incidentally gave me great entertainment in your description of the heronry at Cressi-hall; which is a curiosity I could never manage to see. Four-score nests of such a bird on one tree is a rarity which I would ride half as many miles to have a sight of. Pray be sure to tell me in your next whose seat Cressi-hall is, and near what town it lies.† I have often thought that those vast extents of fens have never been sufficiently explored. If half a dozen gentlemen, furnished with a good strength of water-spaniels, were to beat them over for a week, they would certainly find more species.

There is no bird, I believe, whose manners I have studied more than that of the *caprimulgus* (the goat-sucker),[62] as it is a

* James, chap. iii. 7.
† Cressi-hall is near Spalding, in Lincolnshire.

wonderful and curious creature: but I have always found that though sometimes it may chatter as it flies, as I know it does, yet in general it utters its jarring note sitting on a bough; and I have for many an half hour watched it as it sat with its under mandible quivering, and particularly this summer. It perches usually on a bare twig, with its head lower than its tail, in an attitude well expressed by your draughtsman in the folio *British Zoology*. This bird is most punctual in beginning its song exactly at the close of day; so exactly that I have known it strike up more than once or twice just at the report of the Portsmouth evening gun, which we can hear when the weather is still. It appears to me past all doubt that its notes are formed by organic impulse, by the powers of the parts of its windpipe, formed for sound, just as cats pur. You will credit me, I hope, when I tell you that, as my neighbours were assembled in an hermitage on the side of a steep hill where we drink tea, one of these churn-owls came and settled on the cross of that little straw edifice and began to chatter, and continued his note for many minutes: and we were all struck with wonder to find that the organs of that little animal, when put in motion, gave a sensible vibration to the whole building! This bird also sometimes makes a small squeak, repeated four or five times; and I have observed that to happen when the cock has been pursuing the hen in a toying way through the boughs of a tree.

It would not be at all strange if your bat, which you have procured, should prove a new one, since five species have been found in a neighbouring kingdom. The great sort that I mentioned is certainly a nondescript: I saw but one this summer, and that I had no opportunity of taking.

Your account of the Indian-grass was entertaining. I am no angler myself; but inquiring of those that are, what they supposed that part of their tackle to be made of? they replied 'of the intestines of a silkworm.'

Though I must not pretend to great skill in entomology, yet I cannot say that I am ignorant of that kind of knowledge: I may now and then, perhaps, be able to furnish you with a little information.

The vast rains ceased with us much about the same time as with you, and since we have had delicate weather. Mr Barker, who has measured the rain for more than thirty years, says, in a late letter, that more has fallen this year than in any he ever attended to; though, from July 1763 to January 1764, more fell than in any seven months of this year.

LETTER XXIII

Selborne, February 28, 1769.

DEAR SIR,

It is not improbable that the Guernsey lizard and our green lizard may be specifically the same; all that I know is, that, when some years ago many Guernsey lizards were turned loose in Pembroke college garden, in the University of Oxford, they lived a great while, and seemed to enjoy themselves very well, but never bred. Whether this circumstance will prove anything either way I shall not pretend to say.

I return you thanks for your account of Cressi-hall; but recollect, not without regret, that in June 1746 I was visiting for a week together at Spalding, without ever being told that such a curiosity was just at hand. Pray send me word in your next what sort of tree it is that contains such a quantity of herons' nests; and whether the heronry consists of a whole grove or wood, or only of a few trees.

It gave me satisfaction to find that we accorded so well about the *caprimulgus*: all I contended for was to prove that it often chatters sitting as well as flying; and therefore the noise was voluntary, and from organic impulse, and not from the resistance of the air against the hollow of its mouth and throat.

If ever I saw anything like actual migration, it was last

Michaelmas-day. I was travelling, and out early in the morning: at first there was a vast fog; but, by the time that I was got seven or eight miles from home towards the coast, the sun broke out into a delicate warm day. We were then on a large heath or common, and I could discern, as the mist began to break away, great numbers of swallows (*hirundines rusticæ*) clustering on the stunted shrubs and bushes, as if they had roosted there all night. As soon as the air became clear and pleasant they all were on the wing at once; and, by a placid and easy flight, proceeded on southward towards the sea: after this I did not see any more flocks, only now and then a straggler.

I cannot agree with those persons that assert that the swallow kind disappear some and some gradually, as they come, for the bulk of them seem to withdraw at once: only some stragglers stay behind a long while, and do never, there is the greatest reason to believe, leave this island. Swallows seem to lay themselves up, and to come forth in a warm day, as bats do continually of a warm evening, after they have disappeared for weeks. For a very respectable gentleman assured me that, as he was walking with some friends under Merton-wall on a remarkably hot noon, either in the last week in December or the first week in January, he espied three or four swallows huddled together on the moulding of one of the windows of that college. I have frequently remarked that swallows are seen later at Oxford than elsewhere: is it owing to the vast massy buildings of that place, to the many waters round it, or to what else?

When I used to rise in a morning last autumn, and see the swallows and martins clustering on the chimnies and thatch of the neighbouring cottages, I could not help being touched with a secret delight, mixed with some degree of mortification: with delight to observe with how much ardour and punctuality those poor little birds obeyed the strong impulse towards migration, or hiding, imprinted on their minds by their great Creator; and with some degree of mortification, when I reflected that, after all our pains and inquiries, we are yet not quite certain to what regions they do migrate; and are still farther embarrassed to find that some do not actually migrate at all.

These reflections made so strong an impression on my imagination, that they became productive of a composition that may perhaps amuse you for a quarter of an hour when next I have the honour of writing to you.

LETTER XXIV

Selborne, May 29, 1769.

DEAR SIR,

The *scarabæus fullo*[63] I know very well, having seen it in collections; but have never been able to discover one wild in its natural state. Mr Banks[64] told me he thought it might be found on the sea-coast.

On the thirteenth of April I went to the sheep-down, where the ring-ousels have been observed to make their appearance at spring and fall, in their way perhaps to the north or south; and was much pleased to see three birds about the usual spot. We shot a cock and a hen; they were plump and in high condition. The hen had but very small rudiments of eggs within her, which proves they are late breeders; whereas those species of the thrush kind that remain with us the whole year have fledged young before that time. In their crops was nothing very distinguishable, but somewhat that seemed like blades of vegetables nearly digested. In autumn they feed on haws and yew-berries, and in the spring on ivy-berries. I dressed one of these birds, and found it juicy and well-flavoured. It is remarkable that they make but a few days' stay in their spring visit, but rest near a fortnight at Michaelmas. These birds, from the observations of three springs and two autumns, are most punctual in their return; and exhibit a new migration unnoticed by the writers, who supposed they never were to be seen in any of the southern counties.

One of my neighbours lately brought me a new *salicaria*,[65] which at first I suspected might have proved your willow-lark * but, on a nicer examination, it answered much better to the description of that species which you shot at Revesby, in Lincolnshire. My bird I describe thus: 'It is a size less than the grasshopper-lark; the head, back, and coverts of the wings of a dusky brown, without those dark spots of the grasshopper-lark; over each eye is a milk-white stroke; the chin and throat are white, and the under parts of a yellowish white; the rump is tawny and the feathers of the tail sharp-pointed; the bill is dusky and sharp, and the legs are dusky; the hinder claw long and crooked'. The person that shot it says that it sung so like a reed-sparrow[66] that he took it for one; and that it sings all night; but this account merits further inquiry. For my part, I suspect it is a second sort of *locustella*, hinted at by Dr Derham in Ray's *Letters*: see p. 108. He also procured me a grasshopper-lark.

The question that you put with regard to those genera of animals that are peculiar to America, viz. how they came there, and whence? is too puzzling for me to answer; and yet so obvious as often to have struck me with wonder. If one looks into the writers on that subject little satisfaction is to be found. Ingenious men will readily advance plausible arguments to support whatever theory they shall choose to maintain; but then the misfortune is, every one's hypothesis is each as good as another's, since they are all founded on conjecture. The late writers of this sort, in whom may be seen all the arguments of those that have gone before, as I remember, stock America from the western coast of Africa and the south of Europe; and then break down the Isthmus that bridged over the Atlantic. But this is making use of a violent piece of machinery: it is a difficulty worthy of the interposition of a god! '*Incredulus odi.*'

*For this *salicaria* see letter August 30, 1769.

THE NATURALIST'S SUMMER-EVENING
WALK

> ... equidem credo, quia sit divinitus illis
> Ingenium.   VIRG. *Georg.*

WHEN day declining sheds a milder gleam,
What time the may-fly* haunts the pool or stream;
When the still owl skims round the grassy mead,
What time the timorous hare limps forth to feed;
Then be the time to steal adown the vale,
And listen to the vagrant† cuckoo's tale;
To hear the clamorous‡ curlew call his mate,
Or the soft quail his tender pain relate;
To see the swallow sweep the dark'ning plain
Belated, to support her infant train;
To mark the swift in rapid giddy ring
Dash round the steeple, unsubdu'd of wing:
Amusive birds! – say where your hid retreat
When the frost rages and the tempests beat;
Whence your return, by such nice instinct led,
When spring, soft season, lifts her bloomy head?
Such baffled searches mock man's prying pride,
The GOD of NATURE is your secret guide!

---

*The angler's may-fly, the *ephemera vulgata Linn.*, comes forth from its aurelia state, and emerges out of the water about six in the evening, and dies about eleven at night, determining the date of its fly state in about five or six hours. They usually begin to appear about the 4th of June, and continue in succession for bear a fortnight. See Swammerdam, Derham, Scopoli, etc.

†Vagrant cuckoo; so called because, being tied down by no incubation or attendance about the nutrition of its young, it wanders without control.

‡*Charadrius ædicnemus.*

While deep'ning shades obscure the face of day
To yonder bench, leaf-shelter'd, let us stray,
Till blended objects fail the swimming sight,
And all the fading landscape sinks in night;
To hear the drowsy dor come brushing by
With buzzing wing, or the shrill* cricket cry;
To see the feeding bat glance through the wood;
To catch the distant falling of the flood;
While o'er the cliff th' awakened churn-owl hung
Through the still gloom protracts his chattering song;
While high in air, and pois'd upon his wings,
Unseen, the soft enamour'd woodlark† sings:
These, NATURE's works, the curious mind employ,
Inspire a soothing melancholy joy:
As fancy warms, a pleasing kind of pain
Steals o'er the cheek, and thrills the creeping vein!
Each rural sight, each sound, each smell combine;
The tinkling sheep-bell, or the breath of kine;
The new-mown hay that scents the swelling breeze,
Or cottage-chimney smoking through the trees.
The chilling night-dews fall: away, retire;
For see, the glow-worm lights her amorous fire!‡
Thus, ere night's veil had half obscured the sky,
Th' impatient damsel hung her lamp on high:
True to the signal, by love's meteor led,
Leander hasten'd to his Hero's bed.||

I am, etc.

*Gryllus campestris.
†In hot summer   ts woodlarks soar to a prodigious height, and
hang singing in the air.
‡The light of the female glow-worm (as she often crawls up the stalk
of a grass to make herself more conspicuous) is a signal to the male,
which is a slender dusky scarabæus.
||See the story of Hero and Leander.

LETTER XXV

Selborne, Aug. 30, 1769.

DEAR SIR,

It gives me satisfaction to find that my account of the ousel migration pleases you. You put a very shrewd question when you ask me how I know that their autumnal migration is southward? Was not candour and openness the very life of natural history, I should pass over this query just as the sly commentator does over a crabbed passage in a classic; but common ingenuousness obliges me to confess, not without some degree of shame, that I only reasoned in that case from analogy. For as all other autumnal birds migrate from the northward to us, to partake of our milder winters, and return to the northward again when the rigorous cold abates, so I concluded that the ring-ousels did the same, as well as their congeners the fieldfares; and especially as ring-ousels are known to haunt cold mountainous countries: but I have good reason to suspect since that they may come to us from westward; because I hear, from very good authority, that they breed on Dartmoor; and that they forsake that wild district about the time that our visitors appear, and do not return till late in the spring.

I have taken a great deal of pains about your *salicaria* and mine, with a white stroke over its eye, and a tawny rump. I have surveyed it alive and dead, and have procured several specimens; and am perfectly persuaded myself (and trust you will soon be convinced of the same) that it is no more nor less than the *passer arundinaceus minor* of Ray. This bird, by some means or other, seems to be entirely omitted in the *British Zoology*; and one reason probably was because it is so strangely classed in Ray, who ranges it among his *picis affines*. It ought no doubt to have gone among his *aviculæ caudâ unicolore*, and among your slender-billed small birds of the same division. Linnæus might with great propriety have put it into his genus of *motacilla*; and the *motacilla salicaria* of his *Fauna Suecica* seems to come the nearest to it. It is no uncommon bird, haunting the sides of ponds and rivers where there is covert, and the reeds and sedges of moors. The country people in some places call it the sedge-bird. It sings incessantly night and day during the breeding-time, imitating the note of a sparrow, a swallow, a sky-lark; and has a strange hurrying manner in its song. My specimens correspond most minutely to the description of your fen *salicaria*, shot near Revesby. Mr Ray has given an excellent characteristic of it when he says, '*Rostrum & pedes in hâc aviculâ multò majores sunt quàm pro corporis ratione.*' See letter May 29, 1769.

I have got you the egg of an *œdicnemus*, or stone curlew, which was picked up in a fallow on the naked ground: there were two; but the finder inadvertently crushed one with his foot before he saw them.

When I wrote to you last year on reptiles, I wish I had not forgot to mention the faculty that snakes have of stinking *se defendendo*. I knew a gentleman who kept a tame snake, which was in its person as sweet as any animal while in a good humour and unalarmed; but as soon as a stranger or a dog or cat, came in, it fell to hissing, and filled the room with such nauseous effluvia as rendered it hardly supportable. Thus the squnck, or stonck, of Ray's *Synop. Quadr.* is an innocuous and sweet animal; but, when pressed hard by dogs and men, it can eject such a pestilent and fetid smell and excrement, that nothing can be more horrible.

A gentleman sent me lately a fine specimen of the *lanius minor cinerascens cum maculâ in scapulis albâ Raii*;[67] which is a bird that, at the time of your publishing your two first volumes of *British Zoology*, I find you had not seen. You have described it well from Edwards's drawing.

LETTER XXVI

Selborne, December 8, 1769.

DEAR SIR,

I was much gratified by your communicative letter on your return from Scotland, where you spent, I find, some considerable time, and gave yourself good room to examine the natural curiosities of that extensive kingdom, both those of the islands, as well as those of the highlands. The usual bane of such expeditions is hurry; because men seldom allot themselves half the time they should do: but, fixing on a day for their return, post from place to place, rather as if they were on a journey that required dispatch, than as philosophers investigating the works of nature. You must have made, no doubt, many discoveries, and laid up a good fund of materials for a future edition of the *British Zoology*; and will have no reason to repent that you have bestowed so much pains on a part of Great Britain that perhaps was never so well examined before.

It has always been matter of wonder to me that field-fares, which are so congenerous to thrushes and blackbirds, should never choose to breed in England:[68] but that they should not think even the highlands cold and northerly, and sequestered enough, is a circumstance still more strange and wonderful. The ring-ousel, you find, stays in Scotland the whole year round; so that we have reason to conclude that those migrators that visit us for a short space every autumn do not come from thence.

And here, I think, will be the proper place to mention that those birds were most punctual again in their migration this autumn, appearing, as before, about the 30th of September: but their flocks were larger than common, and their stay protracted somewhat beyond the usual time. If they came to spend the whole winter with us, as some of their congeners do, and then

left us, as they do, in spring, I should not be so much struck
with the occurrence, since it would be similar to that of the
other winter birds of passage; but when I see them for a fortnight at Michaelmas, and again for about a week in the middle
of April, I am seized with wonder, and long to be informed
whence these travellers come, and whither they go, since they
seem to use our hills merely as an inn or baiting place.

Your account of the greater brambling, or snow-fleck, is
very amusing; and strange it is that such a short-winged bird
should delight in such perilous voyages over the northern
ocean! Some country people in the winter time have every
now and then told me that they have seen two or three white
larks on our downs; but on considering the matter, I begin to
suspect that these are some stragglers of the birds we are talking
of, which sometimes perhaps may rove so far to the southward.

It pleases me to find that white hares are so frequent on the
Scottish mountains, and especially as you inform me that it is a
distinct species; for the quadrupeds of Britain are so few, that
every new species is a great acquisition.

The eagle-owl,[69] could it be proved to belong to us, is so
majestic a bird that it would grace our fauna much. I never was
informed before where wild-geese are known to breed.

You admit, I find, that I have proved your fen *salicaria* to be
the lesser reed-sparrow of Ray; and I think that you may be
secure that I am right; for I took very particular pains to clear
up that matter, and had some fair specimens; but, as they were
not well preserved, they are decayed already. You will, no
doubt, insert it in its proper place in your next edition. Your
additional plates will much improve your work.

De Buffon, I know, has described the water shrew-mouse:
but still I am pleased to find you have discovered it in Lincolnshire, for the reason I have given in the article on the white hare.

As a neighbour was lately ploughing in a dry chalky field,
far removed from any water, he turned out a water rat, that
was curiously laid up in an *hybernaculum* artificially formed of

grass and leaves. At one end of the burrow lay above a gallon of potatoes regularly stowed, on which it was to have supported itself for the winter.[70] But the difficulty with me is how this *amphibius mus* came to fix its winter station at such a distance from the water. Was it determined in its choice of that place by the mere accident of finding the potatoes which were planted there; or is it the constant practice of the aquatic rat to forsake the neighbourhood of the water in the colder months?

Though I delight very little in analogous reasoning, knowing how fallacious it is with respect to natural history; yet, in the following instance, I cannot help being inclined to think it may conduce towards the explanation of a difficulty that I have mentioned before, with respect to the invariable early retreat of the *hirundo apus*, or swift, so many weeks before its congeners; and that not only with us, but also in Andalusia, where they also begin to retire about the beginning of August.

The great large bat*[71] (which by the by is at present a nondescript in England, and what I have never been able yet to procure) retires and migrates very early in the summer: it also ranges very high for its food, feeding in a different region of the air; and that is the reason I never could procure one. Now this is exactly the case with the swifts; for they take their food in a more exalted region than the other species, and are very seldom seen hawking for flies near the ground, or over the surface of the water. From hence I would conclude that these *hirundines*, and the larger bats, are supported by some sorts of high-flying gnats, scarabs, or *phalænæ*, that are of short continuance; and that the short stay of these strangers is regulated by the defect of their food.

By my journal it appears that curlews[72] clamoured on to October the thirty-first; since which I have not seen or heard any. Swallows were observed on to November the third.

*The little bat appears almost every month in the year; but I have never seen the large ones till the end of April, nor after July. They are most common in June, but never in any plenty; are a rare species with us.

LETTER XXVII

Selborne, Feb. 22, 1770.

DEAR SIR,

Hedge-hogs abound in my gardens and fields. The manner in which they eat their roots of the plantain in my grass-walks is very curious: with their upper mandible, which is much longer than their lower, they bore under the plant, and so eat the root off upwards, leaving the tuft of leaves untouched. In this respect they are serviceable, as they destroy a very troublesome weed; but they deface the walks in some measure by digging little round holes. It appears, by the dung that they drop upon the turf, that beetles are no inconsiderable part of their food. In June last I procured a litter of four or five young hedge-hogs, which appeared to be about five or six days old; they, I find, like puppies, are born blind, and could not see when they came to my hands. No doubt their spines are soft and flexible at the time of their birth, or else the poor dam would have but a bad time of it in the critical moment of parturition: but it is plain that they soon harden; for these little pigs had such stiff prickles on their backs and sides as would easily have fetched blood, had they not been handled with caution. Their spines are quite white at this age; and they have little hanging ears, which I do not remember to be discernible in the old ones. They can, in part, at this age draw their skin down over their faces; but are not able to contract themselves into a ball as they do, for the sake of defence, when full grown. The reason, I suppose, is, because the curious muscle that enables the creature to roll itself up into a ball was not then arrived at its full tone and firmness. Hedge-hogs make a deep and warm *hybernaculum* with leaves and moss, in which they conceal themselves for the winter: but I never could find that they stored in any winter provision, as some quadrupeds certainly do.

I have discovered an anecdote with respect to the field-fare (*turdus pilaris*), which I think is particular enough: this bird, though it sits on trees in the day-time, and procures the greatest part of its food from white-thorn hedges; yea, moreover, builds on very high trees; as may be seen by the *Fauna Suecica*; yet always appears with us to roost on the ground. They are seen to come in flocks just before it is dark, and to settle and nestle among the heath on our forest. And besides, the larkers, in dragging their nets by night, frequently catch them in the wheat-stubbles; while the bat-fowlers, who take many red-wings in the hedges, never entangle any of this species. Why these birds, in the matter of roosting, should differ from all their congeners, and from themselves also with respect to their proceedings by day, is a fact for which I am by no means able to account.

I have somewhat to inform you of concerning the moose-deer; but in general foreign animals fall seldom in my way; my little intelligence is confined to the narrow sphere of my own observations at home.

LETTER XXVIII

Selborne, March, 1770.

ON Michaelmas-day 1768 I managed to get a sight of the female moose belonging to the Duke of Richmond, at Good-wood; but was greatly disappointed, when I arrived at the spot, to find that it died, after having appeared in a languishing way for some time, on the morning before. However, under-standing that it was not stripped, I proceeded to examine this rare quadruped: I found it in an old green-house, slung under the belly and chin by ropes, and in a standing posture; but, though it had been dead for so short a time, it was in so putrid a state that the stench was hardly supportable. The grand distinction between this deer, and any other species that I have ever met with, consisted in the strange length of its legs; on which it was tilted up much in the manner of birds of the *grallæ* order. I measured it, as they do an horse, and found that, from the ground to the wither, it was just five feet four inches; which height answers exactly to sixteen hands, a growth that few horses arrive at: but then, with this length of legs, its neck was remarkably short, no more than twelve inches; so that, by straddling with one foot forward and the other backward, it grazed on the plain ground, with the greatest difficulty, between its legs: the ears were vast and lopping, and as long as the neck; the head was about twenty inches long, and ass-like; and had such a redundancy of upper lip as I never saw before, with huge nostrils. This lip, travellers say, is esteemed a dainty dish in North America. It is very reasonable to suppose that this creature supports itself chiefly by browsing of trees, and by wading after water-plants; towards which way of livelihood the length of leg and great lip must contribute much. I have read somewhere that it delights in eating the *nymphæa*, or water-lily. From the fore-feet to the belly behind the shoulder it measured three feet and eight inches: the length of the legs before and behind consisted a great deal in the *tibia*, which was strangely long; but in my haste to get out of the stench, I forgot to measure that joint exactly. Its scut seemed to be about an inch long; the colour was a grizzly black; the mane about four inches long; the fore-hoofs were upright and shapely, the hind flat and splayed. The spring before it was only two years old, so that most probably it was not then come to its growth. What a vast tall beast must a full-grown stag be! I have been told some arrive at ten feet and an half! This poor creature had at first a female companion of the same species, which died the spring before. In the same garden was a young stag, or red deer, between whom and this moose it was hoped that there might have been a breed; but their inequality of height must have always been a bar to any commerce of the amorous kind. I should have been glad to have examined the teeth, tongue, lips, hoofs, etc., minutely; but the putrefaction precluded all further curiosity. This animal, the keeper told me, seemed to enjoy itself best in the extreme frost of the former winter. In the house they showed me the horn of a male moose, which had no front-antlers, but only a broad palm with some snags on the edge. The noble owner of the dead moose proposed to make a skeleton of her bones.

Please to let me hear if my female moose corresponds with that you saw; and whether you think still that the American moose and European elk are the same creature.[73] I am,

With the greatest esteem, etc.

LETTER XXIX

Selborne, May 12, 1770.

DEAR SIR,

Last month we had such a series of cold turbulent weather, such a constant succession of frost, and snow, and hail, and tempest, that the regular migration or appearance of the summer birds was much interrupted. Some did not show themselves (at least were not heard) till weeks after their usual time; as the black-cap and white-throat; and some have not been heard yet, as the grasshopper-lark and largest willow-wren. As to the fly-catcher, I have not seen it; it is indeed one of the latest, but should appear about this time: and yet, amidst all this meteorous strife and war of the elements, two swallows discovered themselves as long ago as the eleventh of April, in frost and snow; but they withdrew quickly, and were not visible again for many days. House-martins, which are always more backward than swallows, were not observed till May came in.

Among the monogamous birds several are to be found, after

pairing-time, single, and of each sex: but whether this state of celibacy is matter of choice or necessity, is not so easily discoverable. When the house-sparrows deprive my martins of their nests, as soon as I cause one to be shot, the other, be it cock or hen, presently procures a mate, and so for several times following.

I have known a dove-house infested by a pair of white owls, which made great havoc among the young pigeons: one of the owls was shot as soon as possible; but the survivor readily found a mate, and the mischief went on. After some time the new pair were both destroyed, and the annoyance ceased.

Another instance I remember of a sportsman, whose zeal for the increase of his game being greater than his humanity, after pairing-time he always shot the cock-bird of every couple of partridges upon his grounds; supposing that the rivalry of many males interrupted the breed: he used to say, that, though he had widowed the same hen several times, yet he found she was still provided with a fresh paramour, that did not take her away from her usual haunt.

Again; I knew a lover of setting, an old sportsman, who has often told me that soon after harvest he has frequently taken small coveys of partridges, consisting of cock-birds alone; these he pleasantly used to call old bachelors.

There is a propensity belonging to common house-cats that is very remarkable; I mean their violent fondness for fish, which appears to be their most favourite food: and yet nature in this instance seems to have planted in them an appetite that, unassisted, they know not how to gratify: for of all quadrupeds cats are the least disposed towards water; and will not, when they can avoid it, deign to wet a foot, much less to plunge into that element.

Quadrupeds that prey on fish are amphibious: such is the otter, which by nature is so well formed for diving, that it makes great havoc among the inhabitants of the waters. Not supposing that we had any of those beasts in our shallow brooks,

I was much pleased to see a male otter brought to me, weighing twenty-one pounds, that had been shot on the bank of our stream below the Priory, where the rivulet divides the parish of Selborne from Harteley-wood.

LETTER XXX

Selborne, Aug. 1, 1770.

DEAR SIR,

The French, I think, in general, are strangely prolix in their natural history. What Linnæus says with respect to insects holds good in every other branch: '*Verbositas præsentis sæculi, calamitas artis.*'

Pray how do you approve of Scopoli's new work? As I admire his *Entomologia*, I long to see it.

I forgot to mention in my last letter (and had not room to insert in the former) that the male moose, in rutting time, swims from island to island, in the lakes and rivers of North America, in pursuit of the females. My friend, the chaplain, saw one killed in the water as it was on that errand in the river St Lawrence: it was a monstrous beast, he told me; but he did not take the dimensions.

When I was last in town our friend Mr Barrington most obligingly carried me to see many curious sights. As you were then writing to him about horns, he carried me to see many strange and wonderful specimens. There is, I remember, at Lord Pembroke's, at Wilton, an horn room furnished with more than thirty different pairs; but I have not seen that house lately.

Mr Barrington showed me many astonishing collections of stuffed and living birds from all quarters of the world. After I had studied over the latter for a time, I remarked that every species almost that came from distant regions, such as South

America, the coast of Guinea, etc., were thick-billed birds of the *loxia* and *fringilla* genera; and no *motacillæ*, or *muscicapæ*, were to be met with. When I came to consider, the reason was obvious enough; for the hard-billed birds subsist on seeds, which are easily carried on board; while the soft-billed birds, which are supported by worms and insects, or, what is a *succedaneum* for them, fresh raw meat, can meet with neither in long and tedious voyages. It is from this defect of food that our collections (curious as they are) are defective, and we are deprived of some of the most delicate and lively genera.

I am, etc.

LETTER XXXI

Selborne, Sept. 14, 1770.

DEAR SIR,

You saw, I find, the ring-ousels again among their native crags; and are farther assured that they continue resident in those cold regions the whole year. From whence, then, do our ring-ousels migrate so regularly every September, and make their appearance again, as if in their return, every April? They are more early this year than common, for some were seen at the usual hill on the fourth of this month.

An observing Devonshire gentleman tells me that they frequent some parts of Dartmoor, and breed there; but leave those haunts about the end of September or beginning of October, and return again about the end of March.

Another intelligent person assures me that they breed in great abundance all over the Peak of Derby, and are called there tor-ousels; withdraw in October and November, and return in spring. This information seems to throw some light on my new migration.

Scopoli's* new work (which I have just procured) has its merits in ascertaining many of the birds of the Tirol and Carniola. Monographers, come from whence they may, have, I think, fair pretence to challenge some regard and approbation from the lovers of natural history; for, as no man can alone investigate all the works of nature, these partial writers may, each in their department, be more accurate in their discoveries, and freer from errors, than more general writers; and so by degrees may pave the way to an universal correct natural history. Not that Scopoli is so circumstantial and attentive to the life and conversation of his birds as I could wish: he advances some false facts; as when he says of the *hirundo urbica* that '*pullos extra nidum non nutrit*.' This assertion I know to be wrong from repeated observations this summer, for house-martins do feed their young flying, though it must be acknowledged not so commonly as the house-swallow; and the feat is done in so quick a manner as not to be perceptible to indifferent observers. He also advances some (I was going to say) improbable facts; as when he says of the woodcock that, '*pullos rostro portat fugiens ab hoste.*'[74] But candour forbids me to say absolutely that any fact is false, because I have never been witness to such a fact. I have only to remark that the long unwieldy bill of the woodcock is perhaps the worst adapted of any among the winged creation for such a feat of natural affection.

I am, etc.

LETTER XXXII

Selborne, October 29, 1770.

DEAR SIR,

After an ineffectual search in Linnæus, Brisson, etc., I begin to suspect that I discern my brother's *hirundo hyberna* in Scopoli's new discovered *hirundo rupestris*, p. 167.[75] His des-

* *Annus Primus Historico-Naturalis.*

cription of '*Supra murina, subtus albida; rectrices maculâ ovali albâ in latere interno; pedes nudi, nigri; rostrum nigrum; remiges obscuriores quam plumæ dorsales; rectrices remigibus concolores; caudâ emarginatâ, nec forcipatâ,*' agrees very well with the bird in question; but when he comes to advance that it is '*statura hirundinis urbicæ,*' and that '*definitio hirundinis ripariæ Linnæi huic quoque convenit,*' he in some measure invalidates all he has said; at least he shows at once that he compares them to these species merely from memory: for I have compared the birds themselves, and find they differ widely in every circumstance of shape, size, and colour. However, as you will have a specimen, I shall be glad to hear what your judgment is in the matter.

Whether my brother is forestalled in his nondescript or not, he will have the credit of first discovering that they spend their winters under the warm and sheltery shores of Gibraltar and Barbary.

Scopoli's characters of his *ordines* and *genera* are clear, just, and expressive, and much in the spirit of Linnæus. These few remarks are the result of my first perusal of Scopoli's *Annus Primus*.

The bane of our science is the comparing one animal to the other by memory: for want of caution in this particular, Scopoli falls into errors: he is not so full with regard to the manners of his indigenous birds as might be wished, as you justly observe: his Latin is easy, elegant, and expressive, and very superior to Kramer's.*

I am pleased to see that my description of the moose corresponds so well with yours.

I am, etc.

---

* See his *Elenchus vegetabilium et animalium per Austriam inferiorem, etc.*

LETTER XXXIII

Selborne, Nov. 26, 1770.

DEAR SIR,

I was much pleased to see, among the collection of birds from Gibraltar, some of those short-winged English summer birds of passage, concerning whose departure we have made so much inquiry. Now if these birds are found in Andalusia to migrate to and from Barbary, it may easily be supposed that those that come to us may migrate back to the continent, and spend their winters in some of the warmer parts of Europe. This is certain, that many soft-billed birds that come to Gibraltar appear there only in spring and autumn, seeming to advance in pairs towards the northward, for the sake of breeding during the summer months; and retiring in parties and broods towards the south at the decline of the year: so that the rock of Gibraltar is the great rendezvous, and place of observation, from whence they take their departure each way towards Europe or Africa. It is therefore no mean discovery, I think, to find that our small short-winged summer birds of passage are to be seen spring and autumn on the very skirts of Europe; it is a presumptive proof of their emigrations.

Scopoli seems to me to have found the *hirundo melba*,[76] the great Gibraltar swift, in Tirol, without knowing it. For what is his *hirundo alpina* but the afore-mentioned bird in other words? Says he, 'Omnia prioris' (meaning the swift); 'sed pectus album; paulo major priore.' I do not suppose this to be a new species. It is true also of the *melba*, that 'nidificat in excelsis Alpium rupibus.' *Vid. Annum Primum.*

My Sussex friend, a man of observation and good sense, but no naturalist, to whom I applied on account of the stone curlew, *œdicnemus*, sends me the following account: 'In looking over my Naturalist's Journal for the month of April, I find the stone curlews are first mentioned on the seventeenth and eighteenth, which date seems to me rather late. They live with us all the spring and summer and at the beginning of autumn prepare to take leave by getting together in flocks. They seem to me a bird of passage that may travel into some dry hilly country south of us, probably Spain, because of the abundance of sheep-walks in that country; for they spend their summers with us in such districts. This conjecture I hazard, as I have never met with any one that has seen them in England in the winter. I believe they are not fond of going near the water, but feed on earth-worms, that are common on sheep-walks and downs. They breed on fallows and lay-fields abounding with grey mossy flints, which much resemble their young in colour; among which they skulk and conceal themselves. They make no nest, but lay their eggs on the bare ground, producing in common but two at a time. There is reason to think their young run soon after they are hatched; and that the old ones do not feed them, but only lead them about at the time of feeding, which, for the most part, is in the night.' Thus far my friend.

In the manners of this bird you see there is something very analogous to the bustard, whom it also somewhat resembles in aspect and make, and in the structure of its feet.

For a long time I have desired my relation[77] to look out for these birds in Andalusia; and now he writes me word that, for the first time, he saw one dead in the market on the 3rd of September.

When the *œdicnemus* flies it stretches out its legs straight behind, like an heron.

I am, etc.

LETTER XXXIV

Selborne, March 30, 1771.

DEAR SIR,

There is an insect with us, especially on chalky districts, which is very troublesome and teasing all the latter end of the summer, getting into people's skins, especially those of women and children, and raising tumours which itch intolerably. This animal (which we call an harvest-bug)[78] is very minute, scarce discernible to the naked eye; of a bright scarlet colour, and of the genus of *Acarus*. They are to be met with in gardens on kidney-beans, or any legumens; but prevail only in the hot months of summer. Warreners, as some have assured me, are much infested by them on chalky downs; where these insects swarm sometimes to so infinite a degree as to discolour their nets, and to give them a reddish cast, while the men are so bitten as to be thrown into fevers.

There is a small long shining fly[79] in these parts very troublesome to the housewife, by getting into the chimneys, and laying its eggs in the bacon while it is drying: these eggs produce maggots called jumpers, which, harbouring in the gammons and best parts of the hogs, eat down to the bone, and make great waste. This fly I suspect to be a variety of the *musca putris* of Linnæus: it is to be seen in the summer in the farm-kitchens on the bacon-racks and about the mantelpieces, and on the ceilings.

The insect that infests turnips and many crops in the garden (destroying often whole fields while in their seedling leaves) is an animal that wants to be better known. The country people here call it the turnip-fly and black dolphin;[80] but I know it to be one of the *coleoptera*; the 'chrysomela oleracea, saltatoria, femoribus posticis crassissimis.' In very hot summers they abound to an amazing degree, and as you walk in a field or in a garden, make a pattering like rain, by jumping on the leaves of the turnips or cabbages.

There is an *œstrus*, known in these parts to every ploughboy; which, because it is omitted by Linnæus, is also passed over by late writers, and that is the *curvicauda* of old Moufet, mentioned by Derham in his *Physico-theology*, p. 250: an insect worthy of remark for depositing its eggs as it flies in so dexterous a manner on the single hairs of the legs and flanks of grass-horses[81]. But then Derham is mistaken when he advances that his *œstrus* is the parent of that wonderful star-tailed maggot which he mentions afterwards; for more modern entomologists have discovered that singular production to be derived from the egg of the *musca chamæleon*:[82] see Geoffroy, t. 17, f. 4.

A full history of noxious insects hurtful in the field, garden, and house, suggesting all the known and likely means of destroying them, would be allowed by the public to be a most useful and important work. What knowledge there is of this sort lies scattered, and wants to be collected; great improvements would soon follow of course. A knowledge of the properties, œconomy, propagation, and in short of the life and conversation of these animals, is a necessary step to lead us to some method of preventing their depredations.

As far as I am a judge, nothing would recommend entomology more than some neat plates that should well express the generic distinctions of insects according to Linnæus; for I am well assured that many people would study insects, could they set out with a more adequate notion of those distinctions that can be conveyed at first by words alone.

LETTER XXXV

Selborne, 1771.

DEAR SIR,

Happening to make a visit to my neighbour's peacocks, I could not help observing that the trains of those magnificent birds appear by no means to be their tails; those long feathers growing not from their *uropygium*, but all up their backs. A range of short brown stiff feathers, about six inches long, fixed in the *uropygium*, is the real tail, and serves as the fulcrum to prop the train, which is long and top-heavy, when set on end. When the train is up, nothing appears of the bird before but its head and neck; but this would not be the case were those long feathers fixed only in the rump, as may be seen by the turkey-cock when in a strutting attitude. By a strong muscular vibration these birds can make the shafts of their long feathers clatter like the swords of a sword-dancer; they then trample very quick with their feet, and run backwards towards the females.

I should tell you that I have got an uncommon *calculus ægogropila*, taken out of the stomach of a fat ox; it is perfectly round, and about the size of a large Seville orange; such are, I think, usually flat.

LETTER XXXVI

Sept. 1771.

DEAR SIR,

The summer through I have seen but two of that large species of bat which I call *vespertilio altivolans*,[71] from its manner of feeding high in the air: I procured one of them, and found it to be a male; and made no doubt, as they accompanied together, that the other was a female: but, happening in an evening or two to procure the other likewise, I was somewhat disappointed, when it appeared to be also of the same sex. This circumstance, and the great scarcity of this sort, at least in these parts, occasions some suspicions in my mind whether it is really a species, or whether it may not be the male part of the more known species, one of which may supply many females; as is known to be the case in sheep, and some other quadrupeds. But this doubt can only be cleared by a farther examination, and some attention to the sex, of more specimens: all that I know at present is, that my two were amply furnished with the parts of generation, much resembling those of a boar.

In the extent of their wings they measured fourteen inches and an half, and four inches and an half from the nose to the tip of the tail; their heads were large, their nostrils bilobated, their shoulders broad and muscular, and their whole bodies fleshy and plump. Nothing could be more sleek and soft than their fur, which was of a bright chestnut colour; their maws were full of food, but so macerated that the quality could not be distinguished; their livers, kidneys, and hearts were large, and their bowels covered with fat. They weighed each, when entire, full one ounce and one drachm. Within the ear there was somewhat of a peculiar structure[83] that I did not understand perfectly; but refer it to the observation of the curious anatomist. These creatures send forth a very rancid and offensive smell.

LETTER XXXVII

Selborne, 1771.

DEAR SIR,

On the twelfth of July I had a fair opportunity of contemplating the motions of the *caprimulgus*, or fern-owl, as it was playing round a large oak that swarmed with *scarabæi solstitiales*, or fern-chafers. The powers of its wing were wonderful, exceeding, if possible, the various evolutions and quick turns of the swallow genus. But the circumstance that pleased me most was that I saw it distinctly, more than once, put out its short leg while on the wing, and, by a bend of the head, deliver somewhat into its mouth. If it takes any part of its prey with its foot, as I have now the greatest reason to suppose it does these chafers, I no longer wonder at the use of its middle toe, which is curiously furnished with a serrated claw.[84]

Swallows and martins, the bulk of them, I mean, have forsaken us sooner this year than usual; for, on September the twenty-second, they rendezvoused in a neighbour's walnut-tree, where it seemed probable they had taken up their lodging for the night. At the dawn of the day, which was foggy, they arose all together in infinite numbers, occasioning such a rushing from the strokes of their wings against the hazy air, as might be heard to a considerable distance: since that no flock has appeared, only a few stragglers.

Some swifts staid late, till the twenty-second of August – a rare instance! for they usually withdraw within the first week.*

On September the twenty-fourth three or four ring-ousels appeared in my fields for the first time this season: how punctual are these visitors in their autumnal and spring migrations!

LETTER XXXVIII

Selborne, March 15, 1773.

DEAR SIR,

By my journal for last autumn it appears that the house-martins bred very late, and staid very late in these parts; for, on the first of October, I saw young martins in their nests nearly

*See Letter LIII to Mr Barrington.

fledged; and again, on the twenty-first of October, we had at the next house a nest full of young martins just ready to fly; and the old ones were hawking for insects with great alertness. The next morning the brood forsook their nest, and were flying round the village. From this day I never saw one of the swallow kind till November the third; when twenty, or perhaps thirty, house-martins were playing all day long by the side of the hanging wood, and over my fields. Did these small weak birds, some of which were nestlings twelve days ago, shift their quarters at this late season of the year to the other side of the northern tropic? Or rather, is it not more probable that the next church, ruin, chalk-cliff, steep covert, or perhaps sandbank, lake or pool (as a more northern naturalist would say), may become their *hybernaculum*, and afford them a ready and obvious retreat?

We now begin to expect our vernal migration of ring-ousels every week. Persons worthy of credit assure me that ring-ousels were seen at Christmas 1770 in the forest of Bere, on the southern verge of this county. Hence we may conclude that their migrations are only internal, and not extended to the continent southward, if they do at first come at all from the northern parts of this island only, and not from the north of Europe. Come from whence they will, it is plain, from the fearless disregard that they show for men or guns, that they have been little accustomed to places of much resort. Navigators mention that in the Isle of Ascension, and other such desolate districts, birds are so little acquainted with the human form that they settle on men's shoulders; and have no more dread of a sailor than they would have of a goat that was grazing. A young man at Lewes, in Sussex, assured me that about seven years ago ring-ousels abounded so about that town in the autumn that he killed sixteen himself in one afternoon: he added farther, that some had appeared since in every autumn; but he could not find that any had been observed before the season in which he shot so many. I myself have found these birds in little parties in the autumn cantoned all along the Sussex-downs, wherever there were shrubs and bushes, from Chichester to Lewes; particularly in the autumn of 1770.

I am, etc.

LETTER XXXIX

Selborne, Nov. 9, 1773.

DEAR SIR,

As you desire me to send you such observations as may occur, I take the liberty of making the following remarks, that you may, according as you think me right or wrong, admit or reject what I here advance, in your intended new edition of the *British Zoology.*

The osprey was shot about a year ago at Frinsham-pond, a great lake, at about six miles from hence, while it was sitting on the handle of a plough and devouring a fish: it used to precipitate itself into the water, and so take its prey by surprise.

A great ash-coloured butcher-bird[85] was shot last winter in Tisted-park, and a red-backed butcher-bird at Selborne: they are *raræ aves* in this country.

Crows go in pairs the whole year round.

Cornish choughs[86] abound, and breed on Beachy-head and on all the cliffs of the Sussex coast.

The common wild-pigeon, or stock-dove, is a bird of passage[87] in the south of England, seldom appearing till towards the end of November; is usually the latest winter bird of passage. Before our beechen woods were so much destroyed we had myriads of them, reaching in strings for a mile together as they went out in a morning to feed. They leave us early in spring; where do they breed?

The people of Hampshire and Sussex call the missel-bird the storm-cock, because it sings early in the spring in blowing showery weather; its song often commences with the year: with us it builds much in orchards.

A gentleman assures me that he has taken the nests of ring-ousels on Dartmoor: they build in banks on the sides of streams.

Titlarks[88] not only sing sweetly as they sit on trees, but also as they play and toy about on the wing; and particularly while they are descending, and sometimes as they stand on the ground.

Adamson's testimony seems to me to be a very poor evidence that European swallows migrate during our winter to Senegal: he does not talk at all like an ornithologist; and probably saw only the swallows of that country, which I know build within Governor O'Hara's hall against the roof. Had he known European swallows, would he not have mentioned the species?

The house-swallow washes by dropping into the water as it flies: this species appears commonly about a week before the house-martin, and about ten or twelve days before the swift.

In 1772 there were young house-martins in their nest till October the twenty-third.

The swift appears about ten or twelve days later than the house-swallow: viz., about the twenty-fourth or twenty-sixth of April.

Whin-chats[89] and stone-chatters stay with us the whole year.

Some wheat-ears continue with us the winter through.

Wagtails, all sorts, remain with us all the winter.

Bullfinches, when fed on hempseed, often become wholly black.

We have vast flocks of female chaffinches all the winter, with hardly any males among them.

When you say that in breeding-time the cock-snipes make a bleating noise, and I a drumming (perhaps I should have rather said an humming), I suspect we mean the same thing. However, while they are playing about on the wing they certainly make a loud piping with their mouths: but whether that bleating or humming is ventriloquous, or proceeds from the motion of their wings, I cannot say; but this I know, that when this noise happens the bird is always descending, and his wings are violently agitated.

Soon after the lapwings have done breeding they congregate, and, leaving the moors and marshes, betake themselves to downs and sheep-walks.

Two years ago last spring the little auk was found alive and unhurt, but fluttering and unable to rise, in a lane a few miles from Alresford, where there is a great lake: it was kept a while, but died.

I saw young teals taken alive in the ponds of Wolmer-forest in the beginning of July last, along with flappers, or young wild-ducks.

Speaking of the swift, that page says 'its drink the dew'; whereas it should be 'it drinks on the wing'; for all the swallow kind sip their water as they sweep over the face of pools or rivers: like Virgil's bees, they drink flying, '*flumina summa libant.*' In this method of drinking perhaps this genus may be peculiar.

Of the sedge-bird be pleased to say it sings most part of the night; its notes are hurrying, but not unpleasing, and imitative of several birds; as the sparrow, swallow, skylark. When it happens to be silent in the night, by throwing a stone or clod into the bushes where it sits you immediately set it a-singing; or in other words, though it slumbers sometimes, yet as soon as it is awakened it reassumes its song.

LETTER XL

Selborne, Sept. 2, 1774.

DEAR SIR,

Before your letter arrived, and of my own accord, I had been remarking and comparing the tails of the male and female swallow, and this ere any young broods appeared; so that there was no danger of confounding the dams with their *pulli*: and besides, as they were then always in pairs, and busied in the employ of nidification, there could be no room for mistaking the sexes, nor the individuals of different chimnies the one for the other. From all my observations, it constantly appeared that each sex has the long feathers in its tail that give it that forked shape; with this difference, that they are longer in the tail of the male than in that of the female.

Nightingales, when their young first come abroad, and are helpless, make a plaintive and a jarring noise: and also a snapping or cracking, pursuing people along the hedges as they walk: these last sounds seem intended for menace and defiance.

The grasshopper-lark chirps all night in the height of summer.

Swans turn white the second year, and breed the third.

Weasels prey on moles, as appears by their being sometimes caught in mole-traps.

Sparrow-hawks sometimes breed in old crows' nests, and the kestrel in churches and ruins.

There are supposed to be two sorts of eels in the island of Ely. The threads sometimes discovered in eels are perhaps their young: the generation of eels is very dark and mysterious.[90]

Hen-harriers breed on the ground, and seem never to settle on trees.

When red-starts shake their tails they move them horizontally, as dogs do when they fawn: the tail of a wagtail, when in motion, bobs up and down like that of a jaded horse.

Hedge-sparrows have a remarkable flirt with their wings in breeding-time; as soon as frosty mornings come they make a very piping plaintive noise.

Many birds which become silent about Midsummer reassume their notes again in September; as the thrush, blackbird, woodlark, willow-wren, etc.; hence August is by much the most mute month, the spring, summer, and autumn through.

Are birds induced to sing again because the temperament of autumn resembles that of spring?

Linnæus ranges plants geographically; palms inhabit the tropics, grasses the temperate zones, and mosses and lichens the polar circles; no doubt animals may be classed in the same manner with propriety.

House-sparrows build under eaves in the spring; as the weather becomes hotter they get out for coolness, and nest in plum-trees and apple-trees. These birds have been known sometimes to build in rooks' nests, and sometimes in the forks of boughs under rooks' nests.

As my neighbour was housing a rick he observed that his dogs devoured all the little red mice that they could catch, but rejected the common mice: and that his cats ate the common mice, refusing the red.

Red-breasts sing all through the spring, summer and autumn. The reason that they are called autumn songsters is, because in the two first seasons their voices are drowned and lost in the general chorus; in the latter their song becomes distinguishable. Many songsters of the autumn seem to be the young cock red-breasts of that year:[91] notwithstanding the prejudices in their favour, they do much mischief in gardens to the summer-fruits.*

The titmouse, which early in February begins to make two quaint notes, like the whetting of a saw, is the marsh titmouse:[92] the great titmouse sings with three cheerful joyous notes, and begins about the same time.

Wrens sing all the winter through, frost excepted.

House-martins came remarkably late this year both in Hampshire and Devonshire: is this circumstance for or against either hiding or migration?

Most birds drink sipping at intervals; but pigeons take a long continued draught, like quadrupeds.

*They also eat the berries of the ivy, the honeysuckle, and the *euonymus europæus*, or spindle-tree.

Notwithstanding what I have said in a former letter, no grey crows[93] were ever known to breed on Dartmoor: it was my mistake.

The appearance and flying of the *scarabæus solstitialis*, or fern-chafer,[94] commence with the month of July, and cease about the end of it. These scarabs are the constant food of *caprimulgi*, or fern-owls, through that period. They abound on the chalky downs and in some sandy districts, but not in the clays.

In the garden of the Black-bear inn in the town of Reading is a stream or canal running under the stables and out into the fields on the other side of the road; in this water are many carps, which lie rolling about in sight, being fed by travellers, who amuse themselves by tossing them bread: but as soon as the weather grows at all severe these fishes are no longer seen, because they retire under the stables, where they remain till the return of spring. Do they lie in a torpid state? if they do not, how are they supported?

The note of the white-throat, which is continually repeated, and often attended with odd gesticulations on the wing, is harsh and displeasing. These birds seem of a pugnacious disposition; for they sing with an erected crest and attitudes of rivalry and defiance; are shy and wild in breeding-time, avoiding neighbourhoods, and haunting lonely lanes and commons; nay even the very tops of the Sussex-downs, where there are bushes and covert; but in July and August they bring their broods into gardens and orchards, and make great havoc among the summer-fruits.

The black-cap has in common a full, sweet, deep, loud and wild pipe; yet that strain is of short continuance, and his motions are desultory; but when that bird sits calmly and engages in song in earnest, he pours forth very sweet, but inward melody, and expresses great variety of soft and gentle modulations,[24] superior perhaps to those of any of our warblers, the nightingale excepted.

Black-caps mostly haunt orchards and gardens; while they warble their throats are wonderfully distended.

The song of the red-start is superior, though somewhat like that of the white-throat: some birds have a few more notes than others. Sitting very placidly on the top of a tree in a village, the cock sings from morning to night: he affects neighbourhoods, and avoids solitude, and loves to build in orchards and about houses; with us he perches on the vane of a tall maypole.

The fly-catcher is of all our summer birds the most mute and the most familiar: it also appears the last of any. It builds in a vine, or a sweetbriar, against the wall of an house, or in the hole of a wall, or on the end of a beam or plate, and often close to the post of a door where people are going in and out all day long. This bird does not make the least pretension to song,[95] but uses a little inward wailing note when it thinks its young in danger from cats or other annoyances: it breeds but once, and retires early.

Selborne parish alone can and has exhibited at times more than half the birds that are ever seen in all Sweden; the former has produced more than one hundred and twenty species, the latter only two hundred and twenty-one. Let me add also that it has shown near half the species that were ever known in Great Britain.*

On a retrospect, I observe that my long letter carries with it a quaint and magisterial air, and is very sententious: but, when I recollect that you requested stricture and anecdote, I hope you will pardon the didactic manner for the sake of the information it may happen to contain.

*Sweden, 221; Great Britain, 252 species. [96]

LETTER XLI

IT is matter of curious inquiry to trace out how those species of soft-billed birds, that continue with us the winter through, subsist during the dead months. The imbecility[97] of birds seems not to be the only reason why they shun the rigour of our winters; for the robust wryneck (so much resembling the hardy race of wood-peckers) migrates, while the feeble little golden-crowned wren, that shadow of a bird, braves our severest frosts without availing himself of houses or villages, to which most of our winter birds crowd in distressful seasons, while this keeps aloof in fields and woods; but perhaps this may be the reason why they may often perish, and why they are almost as rare as any bird we know.

I have no reason to doubt but that the soft-billed birds, which winter with us, subsist chiefly on insects in their *aurelia* state.[98] All the species of wagtails in severe weather haunt shallow streams near their spring-heads, where they never freeze; and, by wading, pick out the aurelias of the genus of *Phryganeæ*,* etc.

Hedge-sparrows frequent sinks and gutters in hard weather, where they pick up crumbs and other sweepings: and in mild weather they procure worms, which are stirring every month in the year, as any one may see that will only be at the trouble of taking a candle to a grass-plot on any mild winter's night. Red-breasts and wrens in the winter haunt out-houses, stables, and barns, where they find spiders and flies that have laid themselves up during the cold season. But the grand support of the soft-billed birds in winter is that infinite profusion of *aureliæ* of the *lepidoptera ordo*, which is fastened to the twigs of trees and

* See Derham's *Physico-theology*, p. 235.

their trunks; to the pales and walls of gardens and buildings; and is found in every cranny and cleft of rock or rubbish, and even in the ground itself.

Every species of titmouse winters with us; they have what I call a kind of intermediate bill between the hard and the soft, between the Linnæan genera of *fringilla* and *motacilla*. One species alone spends its whole time in the woods and fields, never retreating for succour in the severest seasons to houses and neighbourhoods; and that is the delicate long-tailed titmouse, which is almost as minute as the golden-crowned wren: but the blue titmouse, or nun (*parus cœruleus*), the cole-mouse (*parus ater*), the great black-headed titmouse (*fringillago*), and the marsh titmouse (*parus palustris*), all resort, at times, to buildings; and in hard weather particularly. The great titmouse, driven by stress of weather, much frequents houses, and, in deep snows, I have seen this bird, while it hung with its back downwards (to my no small delight and admiration), draw straw lengthwise from out the eaves of thatched houses, in order to pull out the flies that were concealed between them, and that in such numbers that they quite defaced the thatch, and gave it a ragged appearance.

The blue titmouse, or nun, is a great frequenter of houses, and a general devourer. Beside insects, it is very fond of flesh; for it frequently picks bones on dung-hills: it is a vast admirer of suet, and haunts butchers' shops. When a boy, I have known twenty in a morning caught with snap mousetraps, baited with tallow or suet. It will also pick holes in apples left on the ground, and be well entertained with the seeds on the head of a sunflower. The blue, marsh, and great titmice will, in very severe weather, carry away barley and oat straws from the sides of ricks.

How the wheat-ear and whin-chat support themselves in winter cannot be so easily ascertained, since they spend their time on wild heaths and warrens; the former especially, where there are stone quarries: most probably it is that their maintenance arises from the *aureliæ* of the *lepidoptera ordo*, which furnish them with a plentiful table in the wilderness.

I am, etc.

LETTER XLII

Selborne, March 9, 1775.

DEAR SIR,

Some future faunist, a man of fortune, will, I hope, extend his visits to the kingdom of Ireland; a new field, and a country little known to the naturalist. He will not, it is to be wished, undertake that tour unaccompanied by a botanist, because the mountains have scarcely been sufficiently examined; and the southerly counties of so mild an island may possibly afford some plants little to be expected[99] within the British dominions. A person of a thinking turn of mind will draw many just remarks from the modern improvements of that country, both in arts and agriculture, where premiums obtained long before they were heard of with us. The manners of the wild natives, their superstitions, their prejudices, their sordid way of life, will extort from him many useful reflections. He should also take with him an able draughtsman; for he must by no means pass over the noble castles and seats, the extensive and picturesque lakes and water-falls, and the lofty stupendous mountains, so little known, and so engaging to the imagination when described and exhibited in a lively manner: such a work would be well received.

As I have seen no modern map of Scotland, I cannot pretend to say how accurate or particular any such may be; but this I know, that the best old maps of that kingdom are very defective.

The great obvious defect that I have remarked in all maps of Scotland that have fallen in my way is, a want of a coloured line, or stroke, that shall exactly define the just limits of that district called the Highlands. Moreover, all the great avenues to that mountainous and romantic country want to be well distinguished .The military roads formed by General Wade are so great and Roman-like an undertaking that they well merit attention. My old map, Moll's Map, takes notice of Fort William; but could not mention the other forts that have been erected long since: therefore a good representation of the chain of forts should not be omitted.

The celebrated zigzag up the Coryarich must not be passed over. Moll takes notice of Hamilton and Drumlanrig, and such capital houses; but a new survey, no doubt, should represent every seat and castle remarkable for any great event, or celebrated for its paintings, etc. Lord Breadalbane's seat and beautiful policy are too curious and extraordinary to be omitted.

The seat of the Earl of Eglintoun, near Glasgow, is worthy of notice. The pine plantations of that nobleman are very grand and extensive indeed.

I am, etc.

LETTER XLIII

DEAR SIR,

A pair of honey-buzzards,[100] *buteo apivorus, sive vespivorus Raii,* built them a large shallow nest, composed of twigs and lined with dead beechen leaves, upon a tall slender beech near the middle of Selborne-hanger, in the summer of 1780. In the middle of the month of June a bold boy climbed this tree, though standing on so steep and dizzy a situation, and brought down an egg, the only one in the nest, which had been sat on for some time, and contained the embrio of a young bird. The egg was smaller, and not so round as those of the common buzzard; was dotted at each end with small red spots, and surrounded in the middle with a broad bloody zone.

The hen-bird was shot, and answered exactly to Mr Ray's description of that species; had a black cere, short thick legs, and a long tail. When on the wing this species may be easily distinguished from the common buzzard by its hawk-like appearance, small head, wings not so blunt, and longer tail. This specimen contained in its craw some limbs of frogs, and many grey snails without shells. The irides of the eyes of this bird were of a beautiful bright yellow colour.

About the tenth of July in the same summer a pair of sparrow-hawks bred in an old crow's nest on a low beech in the same hanger; and as their brood, which was numerous, began to grow up, became so daring and ravenous, that they were a terror to all the dames in the village that had chickens or ducklings under their care. A boy climbed the tree, and found the young so fledged that they all escaped from him: but discovered that a good house had been kept: the larder was well-stored with provisions; for he brought down a young blackbird, jay, and house martin, all clean picked, and some half devoured. The old birds had been observed to make sad havoc for some days among the new-flown swallows and martins, which, being but lately out of their nests, had not acquired those powers and command of wing that enable them, when more mature, to set such enemies at defiance.

LETTER XLIV

Selborne, Nov. 30, 1780.

Dear Sir,

Every incident that occasions a renewal of our correspondence will ever be pleasing and agreeable to me.

As to the wild wood-pigeon, the *œnas*, or *vinago*, of Ray, I am much of your mind; and see no reason for making it the origin of the common house-dove: but suppose those that have advanced that opinion may have been misled by another appellation, often given to the *œnas*, which is that of stock-dove.

Unless the stock-dove in the winter varies greatly in manners from itself in summer, no species seems more unlikely to be domesticated, and to make an house-dove. We very rarely see the latter settle on trees at all, nor does it ever haunt the woods; but the former, as long as it stays with us, from November perhaps to February, lives the same wild life with the ring-dove, *palambus torquatus*; frequents coppices and groves, supports itself chiefly by mast, and delights to roost in the tallest beeches. Could it be known in what manner stock-doves build, the doubt would be settled with me at once, provided they construct their nests on trees, like the ring-dove, as I much suspect they do.

You received, you say, last spring a stock-dove from Sussex; and are informed that they sometimes breed in that county. But why did not your correspondent determine the place of its nidification, whether on rocks, cliffs, or trees? If he was not an adroit ornithologist I should doubt the fact, because people with us perpetually confound the stock-dove with the ring-dove.

For my own part, I readily concur with you in supposing that house-doves are derived from the small blue rock-pigeon, for many reasons. In the first place, the wild stock-dove is manifestly larger than the common house-dove, against the usual rule of domestication, which generally enlarges the breed. Again, these two remarkable black spots on the remiges of each wing of the stock-dove, which are so characteristic of the species, would not, one should think, be totally lost by its being reclaimed; but would often break out among its descendants. But what is worth an hundred arguments is, the instance you give in Sir Roger Mostyn's house-doves, in Caernarvonshire; which, though tempted by plenty of food and gentle treatment, can never be prevailed on to inhabit their cote for any time; but as soon as they begin to breed, betake themselves to the fastnesses of Ormshead, and deposit their young in safety amidst the inaccessible caverns and precipices of that stupendous promontory.

Naturam expellas furcâ, tamen usque recurret.

I have consulted a sportsman, now in his seventy-eighth year, who tells me that fifty or sixty years back, when the beechen woods were much more extensive than at present, the number of wood-pigeons was astonishing; that he has often killed near twenty in a day; and that with a long wild-fowl piece he has shot seven or eight at a time on the wing as they came wheeling over his head: he moreover adds, which I was not aware of, that often there were among them little parties of small blue doves, which he calls *rockiers*. The food of these numberless emigrants was beech-mast and some acorns; and particularly barley, which they collected in the stubbles. But of late years, since the vast increase of turnips, that vegetable has furnished a great part of their support in hard weather; and the holes they pick in these roots greatly damage the crop. From this food their flesh has contracted a rancidness which occasions them to be rejected by nicer judges of eating, who thought them before a delicate dish. They were shot not only as they were feeding in the fields, and especially in snowy weather, but also at the close of the evening, by men who lay in ambush among the woods and groves to kill them as they came in to roost.* These are the principal circumstances relating to this wonderful internal migration, which with us takes place towards the end of November, and ceases early in the spring. Last winter we had in Selborne high wood about an hundred of these doves; but in former times the flocks were so vast not only

*Some old sportsmen say that the main part of these flocks used to withdraw as soon as the heavy Christmas frosts were over.

with us but all the district round, that on mornings and evenings they traversed the air, like rooks, in strings, reaching for a mile together. When they thus rendezvoused here by thousands, if they happened to be suddenly roused from their roost-trees on an evening,

> Their rising all at once was like the sound
> Of thunder heard remote. . . .

It will by no means be foreign to the present purpose to add, that I had a relation in this neighbourhood who made it a practice for a time, whenever he could procure the eggs of a ring-dove, to place them under a pair of doves that were sitting in his own pigeon-house; hoping thereby, if he could bring about a coalition, to enlarge his breed, and teach his own doves to beat out into the woods and to support themselves by mast: the plan was plausible, but something always interrupted the success; for though the birds were usually hatched, and sometimes grew to half their size, yet none ever arrived at maturity. I myself have seen these foundlings in their nest displaying a strange ferocity of nature, so as scarcely to bear to be looked at, and snapping with their bills by way of menace. In short, they always died, perhaps for want of proper sustenance: but the owner thought that by their fierce and wild demeanour they frighted their foster-mothers, and so were starved.

Virgil, as a familiar occurrence, by way of simile, describes a dove haunting the cavern of a rock in such engaging numbers that I cannot refrain from quoting the passage: and John Dryden has rendered it so happily in our language, that without farther excuse I shall add his translation also.

> Qualis speluncâ subitò commota Columba,
> Cui domus, et dulces latebroso in pumice nidi,
> Fertur in arva volans, plausumque exterrita pennis
> Dat tecto ingentem – mox aere lapsa quieto,
> Radit iter liquidum, celeres neque commovet alas.

As when a dove her rocky hold forsakes,
Rous'd, in her fright her sounding wings she shakes;
The cavern rings with clattering: out she flies,
And leaves her callow care, and cleaves the skies:
At first she flutters: but at length she springs
To smoother flight, and shoots upon her wings.

I am, etc.

Letters to Daines Barrington

LETTER I

Selborne, June 30, 1769.

DEAR SIR,

When I was in town last month I partly engaged that I would sometime do myself the honour to write to you on the subject of natural history: and I am the more ready to fulfil my promise, because I see you are a gentleman of great candour, and one that will make allowances; especially where the writer professes to be an out-door naturalist, one that takes his observations from the subject itself, and not from the writings of others.

The following is a LIST *of the* SUMMER BIRDS *of* PASSAGE *which I have discovered in this neighbourhood, ranged somewhat in the order in which they appear.*

		RAII NOMINA.	USUALLY APPEARS ABOUT
1.	Wry-neck,	*Jynx, sive torquilla:*	The middle of March: harsh note.
2.	Smallest willow-wren,	*Regulus non cristatus:*	March 23: chirps till September.
3.	Swallow,	*Hirundo domestica:*	April 13.
4.	Martin,	*Hirundo rustica:*	Ditto.
5.	Sand-martin,	*Hirundo riparia:*	Ditto.
6.	Black-cap,	*Atricapilla:*	Ditto: a sweet wild note.
7.	Nightingale,	*Luscinia:*	Beginning of April.
8.	Cuckoo,	*Cuculus:*	Middle of April.
9.	Middle willow-wren,	*Regulus non cristatus:*	Ditto, a sweet plaintive note.
10.	White-throat,	*Ficedulæ affinis:*	Middle of April: mean note; sings on till September.
11.	Red-start,	*Ruticilla:*	Ditto: more agreeable song.

	RAII NOMINA.	USUALLY APPEARS ABOUT
12. Stone curlew,	*Œdicnemus:*	End of March; loud nocturnal whistle.
13. Turtle-dove,	*Turtur:*	
14. Grasshopper-lark,	*Alauda minima locustæ voce*	Middle of April: a small sibilous note, till the end of July.
15. Swift,	*Hirundo apus:*	About April 27.
16. Less reed-sparrow[101],	*Passer arundinaceus minor:*	A sweet polyglot, but hurrying: it has the notes of many birds.
17. Land-rail,	*Ortygometra:*	A loud harsh note, *crex, crex.*
18. Largest willow-wren,	*Regulus non cristatus:*	Cantat voce striduld locustæ; end of April, on the tops of high beeches.
19. Goat-sucker, or fern-owl,	*Caprimulgus:*	Beginning of May; chatters by night with a singular noise.
20. Fly-catcher,	*Stoparola:*	May 12. A very mute bird: this is the latest summer bird of passage.

This assemblage of curious and amusing birds belongs to ten several genera of the Linnæan system; and are all of the *ordo* of *passeres*, save the *jynx* and *cuculus*, which are *picæ*, and the *charadrius* (*œdicnemus*) and *rallus* (*ortygometra*) which are *grallæ*.

These birds, as they stand numerically, belong to the following Linnæan genera:

1.	*Jynx.*	13.	*Columba.*
2, 6, 7, 9, 10, 11, 16, 18. }	*Motacilla.*	17.	*Rallus.*
3, 4, 5, 15.	*Hirundo.*	19.	*Caprimulgus.*
8.	*Cuculus.*	14.	*Alauda.*
12.	*Charadrius.*	20.	*Muscicapa.*

Most soft-billed birds live on insects, and not on grain and seeds; and therefore at the end of summer they retire: but the following soft-billed birds, though insect-eaters, stay with us the year round:

RAII NOMINA.

Red-breast,	*Rubecula:*	These frequent houses; and haunt outbuildings in the winter; eat spiders.
Wren,	*Paser troglodytes:*	
Hedge-sparrow,	*Curruca:*	Haunt sinks for crumbs and other sweepings.
White-wagtail,	*Motacilla alba:*	These frequent shallow rivulets near the spring heads, where they never freeze: eat the aureliæ of *Phryganea.* The smallest birds that walk.
Yellow-wagtail,	*Motacilla flava:*	
Grey-wagtail,	*Motacilla cinerea:*	
Wheat-ear,	*Œnanthe:*	Some of these are to be seen with us the winter through.
Whin-chat,	*Œnanthe secunda:*	
Stone-chatter,	*Œnanthe tertia:*	
Golden-crowned wren,	*Regulus cristatus:*	This is the smallest British bird: haunts the tops of all trees; stays the winter through.

A LIST *of the* WINTER BIRDS *of* PASSAGE *round this neighbourhood, ranged somewhat in the order in which they appear:*

RAII NOMINA.

1. Ring-ousel,	*Merula torquata:*	This is a new migration which I have lately discovered about Michaelmas week, and again about the fourteenth of March.
2. Redwing,	*Turdus iliacus:*	About old Michaelmas.
3. Fieldfare,	*Turdus pilaris:*	Though a percher by day, roosts on the ground.
4. Royston-crow,[102]	*Cornix cinerea:*	Most frequent on downs.
5. Wood-cock,	*Scolopax:*	Appears about old Michaelmas.
6. Snipe,	*Gallinago minor:*	Some snipes constantly breed with us.

RAII NOMINA.

7. Jack-snipe,   *Gallinago minima:*

8. Wood-pigeon,   *Œnas:*   Seldom appears till late: not in such plenty as formerly.

9. Wild-swan,   *Cygnus ferus:*   On some large waters.

10. Wild-goose,   *Anser ferus:*

11. Wild-duck,[103]   *Anas torquata minor:*

12. Pochard,   *Anas fera fusca:*

13. Widgeon,   *Penelope:*   On our streams and lakes.

14. Teal, breeds with us in Wolmer-forest.   *Querquedula:*

15. Gross-beak,   *Coccothraustes:*

16. Cross-bill,   *Loxia:*   These are only wanderers that appear occasionally, and are not observant of any regular migration.

17. Silk-tail,   *Garrulus bohemicus:*

These birds, as they stand numerically, belong to the following Linnæan genera:

1, 2, 3.	*Turdus.*	9, 10, 11, 12, 13, 14.	*Anas.*
4.	*Corvus.*	15, 16.	*Loxia.*
5, 6, 7.	*Scolopax.*	17.	*Ampelis.*
8.	*Columba.*		

Birds that sing in the night are but few:

Nightingale,	*Luscinia:*	'In shadiest covert hid.' – MILTON.
Woodlark,	*Alauda arborea:*	Suspended in mid air.
Less reed-sparrow,	*Passer arundinaceus minor.*	Among reeds and willows.

I should now proceed to such birds as continue to sing after Midsummer, but, as they are rather numerous, they would exceed the bounds of this paper: besides, as this is now the season for remarking on that subject, I am willing to repeat my observations on some birds concerning the continuation of whose song I seem at present to have some doubt.

I am, etc.

LETTER II

Selborne, Nov. 2, 1769.

Dear Sir,

When I did myself the honour to write to you about the end of last June on the subject of natural history, I sent you a list of the summer birds of passage which I have observed in this neighbourhood; and also a list of the winter birds of passage; I mentioned besides those soft-billed birds that stay with us the winter through in the south of England, and those that are remarkable for singing in the night.

According to my proposal, I shall now proceed to such birds (singing birds strictly so called) as continue in full song till after Midsummer; and shall range them somewhat in the order in which they first begin to open as the spring advances.

		RAII NOMINA.	
1.	Woodlark,	*Alauda arborea:*	In January, and continues to sing through all the summer and autumn.
2.	Song-thrush,	*Turdus simpliciter dictus:*	In February and on to August, reassume their song in autumn.
3.	Wren,	*Passer troglodytes:*	All the year, hard frost excepted.
4.	Red-breast,	*Rubecula:*	Ditto.
5.	Hedge-sparrow,	*Curruca:*	Early in February to July the 10th.
6.	Yellow-hammer,	*Emberiza flava:*	Early in February, and on through July to August the 21st.
7.	Skylark,	*Alauda vulgaris:*	In February, and on to October.
8.	Swallow,	*Hirundo domestica:*	From April to September.
9.	Black-cap,	*Atricapilla:*	Beginning of April to July 13th.

RAII NOMINA.

10. Titlark,	*Alauda pratorum:*	From middle of April to July the 16th.
11. Blackbird,	*Merula vulgaris:*	Sometimes in February and March, and so on to July the twenty-third; reassumes in autumn.
12. White-throat,	*Ficedulæ affinis:*	In April and on to July 23.
13. Goldfinch,	*Carduelis:*	April and through to September 16.
14. Greenfinch,	*Chloris:*	On to July and August 2.
15. Less reed-sparrow,	*Passer arundinaceus minor:*	May, on to beginning of July.
16. Common linnet,	*Linaria vulgaris:*	Breeds and whistles on till August; reassumes its note when they begin to congregate in October, and again early before the flock separate.

Birds that cease to be in full song, and are usually silent at or before Midsummer:

17. Middle willow-wren,	*Regulus non cristatus:*	Middle of June: begins in April.
18. Red-start,	*Ruticilla:*	Middle of June: begins in May.
19. Chaffinch,	*Fringilla:*	Beginning of June: sings first in February.
20. Nightingale,	*Luscinia:*	Middle of June: sings first in April.

Birds that sing for a short time, and very early in the spring:

21. Missel-bird,	*Turdus viscivorus:*	January the 2nd, 1770, in February. Is called in Hampshire and Sussex the storm-cock, because its song is supposed to forebode windy wet weather: is the largest singing bird we have.

RAII NOMINA.

22. Great tit-mouse, or ox-eye,	*Fringillago:*	In February, March, April: reassumes for a short time in September.

Birds that have somewhat of a note or song, and yet are hardly to be called singing birds:

23. Golden-crowned wren,	*Regulus cristatus:*	Its note as minute as its person; frequents the tops of high oaks and firs; the smallest British bird.
24. Marsh titmouse,	*Parus palustris:*	Haunts great woods; two harsh sharp notes.
25. Small willow-wren,	*Regulus non cristatus:*	Sings in March and on to September.
26. Largest ditto,	*Ditto:*	*Cantat voce stridulâ locustæ:* from end of April to August.
27. Grasshopper-lark,	*Alauda minima voce locustæ:*	Chirps all night, from the middle of April to the end of July.
28. Martin,	*Hirundo agrestis:*	All the breeding time; from May to September.
29. Bullfinch,	*Pyrrhula:*	
30. Bunting,[104]	*Emberiza alba:*	From the end of January to July.

All singing birds, and those that have any pretensions to song, not only in Britain, but perhaps the world through, come under the Linnæan *ordo* of *passeres.*

The above-mentioned birds, as they stand numerically, belong to the following Linnæan genera.

1, 7, 10, 27.	*Alauda.*	8, 28.	*Hirundo.*
2, 11, 21.	*Turdus.*	13, 16, 19.	*Fringilla.*
3, 4, 5, 9, 12, 15, 17, 18, 20, 23, 25, 26.	} *Motacilla.*	22, 24.	*Parus.*
6, 30.	*Emberiza.*	14, 29.	*Loxia.*

Birds that sing as they fly are but few:

	RAII NOMINA.	
Skylark,	*Alauda vulgaris:*	Rising, suspended, and falling.
Titlark,	*Alauda pratorum:*	In its descent; also sitting on trees, and walking on the ground.
Woodlark,	*Alauda arborea:*	Suspended; in hot summer nights all night long.
Blackbird,	*Merula:*	Sometimes from bush to bush.
White-throat,	*Ficedulæ affinis:*	Uses when singing on the wing odd jerks and gesticulations.
Swallow,	*Hirundo domestica:*	In soft sunny weather.
Wren,	*Passer troglodytes:*	Sometimes from bush to bush.

Birds that breed most early in these parts:

Raven,	*Corvus:*	Hatches in February and March.
Song-thrush,	*Turdus:*	In March.
Blackbird,	*Merula:*	In March.
Rook,	*Cornix frugilega:*	Builds the beginning of March.
Woodlark,	*Alauda arborea:*	Hatches in April.
Ring-dove,	*Palumbus torquatus:*	Lays the beginning of April.

All birds that continue in full song till after Midsummer appear to me to breed more than once.

Most kinds of birds seem to me to be wild and shy somewhat in proportion to their bulk; I mean in this island, where they are much pursued and annoyed: but in Ascension-island, and many other desolate places, mariners have found fowls so unacquainted with an human figure, that they would stand still to be taken; as is the case with boobies, etc. As an example of what is advanced, I remark that the golden-crested wren (the smallest British bird) will stand unconcerned till you come within three or four yards of it, while the bustard (*otis*), the largest British land fowl, does not care to admit a person within so many furlongs.

I am, etc.

LETTER III

Selborne, Jan. 15, 1770.

DEAR SIR,

It was no small matter of satisfaction to me to find that you were not displeased with my little *methodus* of birds. If there was any merit in the sketch, it must be owing to its punctuality. For many months I carried a list in my pocket of the birds that were to be remarked, and, as I rode or walked about my business, I noted each day the continuance or omission of each bird's song; so that I am as sure of the certainty of my facts as a man can be of any transaction whatsoever.

I shall now proceed to answer the several queries which you put in your two obliging letters, in the best manner that I am able. Perhaps Eastwick, and its environs, where you heard so very few birds, is not a woodland country, and therefore not stocked with such songsters. If you will cast your eye on my last letter, you will find that many species continued to warble after the beginning of July.

The titlark and yellowhammer breed late, the latter very late; and therefore it is no wonder that they protract their song; for I lay it down as a maxim in ornithology, that as long as there is any incubation going on there is music. As to the red-breast and wren, it is well known to the most incurious observer that they whistle the year round, hard frost excepted; especially the latter.

It was not in my power to procure you a black-cap, or a less reed-sparrow, or sedge-bird, alive. As the first is undoubtedly, and the last, as far as I can see yet, a summer bird of passage, they would require more nice and curious management in a cage than I should be able to give them; they are both distinguished songsters. The note of the former has such a wild sweetness that it always brings to my mind those lines in a song in *As You Like It*,

> And tune his merry note
> Unto the *wild* bird's throat.
>
> SHAKESPEARE.

The latter has a surprising variety of notes resembling the song of several other birds; but then it also has an hurrying manner, not at all to its advantage; it is notwithstanding a delicate polyglot.

It is new to me that titlarks in cages sing in the night; perhaps only caged birds do so. I once knew a tame redbreast in a cage that always sang as long as candles were in the room; but in their wild state no one supposes they sing in the night.[105]

I should be almost ready to doubt the fact, that there are to be seen much fewer birds in July than in any former month, notwithstanding so many young are hatched daily. Sure I am that it is far otherwise with respect to the swallow tribe, which increases prodigiously as the summer advances: and I saw, at the time mentioned, many hundreds of young wagtails on the banks of the Cherwell, which almost covered the meadows. If the matter appears as you say in the other species, may it not be owing to the dams being engaged in incubation, while the young are concealed by the leaves?

Many times have I had the curiosity to open the stomachs of woodcocks and snipes; but nothing ever occurred that helped to explain to me what their subsistence might be: all that I could ever find was a soft mucus, among which lay many pellucid small gravels.

I am, etc.

LETTER IV

Selborne, Feb. 19, 1770.

DEAR SIR,

Your observation that 'the cuckoo does not deposit its egg indiscriminately in the nest of the first bird that comes in its way, but probably looks out a nurse in some degree congenerous, with whom to intrust its young,' is perfectly new to me; and struck me so forcibly, that I naturally fell into a train of thought that led me to consider whether the fact was so, and what reason there was for it. When I came to recollect and inquire, I could not find that any cuckoo had ever been seen in these parts, except in the nest of the wagtail, the hedge-sparrow, the titlark, the white-throat, and the red-breast, all soft-billed insectivorous birds. The excellent Mr Willughby mentions the nest of the *palumbus* (ring-dove), and of the *fringilla* (chaffinch), birds that subsist on acorns and grains, and such hard food: but then he does not mention them as of his own knowledge: but says afterwards that he saw himself a wagtail feeding a cuckoo. It appears hardly possible that a soft-billed bird should subsist on the same food with the hard-billed:[106] for the former have thin membranaceous stomachs suited to their soft food; while the latter, the granivorous tribe, have strong muscular gizzards, which, like mills, grind, by the help of small gravels and pebbles what is swallowed. This proceeding of the cuckoo, of dropping its eggs as it were by chance, is such a monstrous outrage on maternal affection, one of the first great dictates of nature, and such a violence on instinct, that, had it only been related of a bird in the Brazils, or Peru, it would never have merited our belief. But yet, should it farther appear that this simple bird, when divested of the natural στοργή that seems to raise the kind in general above themselves, and inspire them with extraordinary degrees of cunning and address, may be still endued with a more enlarged faculty of discerning what species are suitable and congenerous nursing-mothers for its disregarded eggs and young, and may deposit them only under their care, this would be adding wonder to wonder, and instancing in a fresh manner that the methods of Providence are not subjected to any mode or rule, but astonish us in new lights, and in various and changeable appearances.

What was said by a very ancient and sublime writer concerning the defect of natural affection in the ostrich, may be well applied to the bird we are talking of:

'She is hardened against her young ones, *as though they were not* hers:
Because God hath deprived her of wisdom, neither hath he imparted. to her understanding.'*

Query. – Does each female cuckoo lay but one egg in a season, or does she drop several in different nests according as opportunity offers?

I am, etc.

LETTER V

Selborne, April 12, 1770.

DEAR SIR,
I heard many birds of several species sing last year after Midsummer; enough to prove that the summer solstice is not the period that puts a stop to the music of the woods. The yellow-hammer no doubt persists with more steadiness than any other; but the woodlark, the wren, the red-breast, the swallow, the white-throat, the goldfinch, the common linnet, are all undoubted instances of the truth of what I advance.

If this severe season does not interrupt the regularity of the summer migrations, the black-cap will be here in two or three

*Job xxxix. 16, 17.

days. I wish it was in my power to procure you one of those songsters; but I am no birdcatcher; and so little used to birds in a cage, that I fear if I had one it would soon die for want of skill in feeding.

Was your reed-sparrow, which you kept in a cage, the thick-billed reed-sparrow[107] of the *Zoology*, p. 320; or was it the less reed-sparrow of Ray, the sedge-bird of Mr Pennant's last publication, p. 16?

As to the matter of long-billed birds growing fatter in moderate frosts, I have no doubt within myself what should be the reason. The thriving at those times appears to me to arise altogether from the gentle check which the cold throws upon insensible perspiration. The case is just the same with black-birds, etc.; and farmers and warreners observe, the first, that their hogs fat more kindly at such times, and the latter that the rabbits are never in such good case as in a gentle frost. But when frosts are severe, and of long continuance, the case is soon altered; for then a want of food soon overbalances the repletion occasioned by a checked perspiration. I have observed, more-over, that some human constitutions are more inclined to plumpness in winter than in summer.

When birds come to suffer by severe frost, I find that the first that fail and die are the redwing-fieldfares, and then the song-thrushes.

You wonder, with good reason, that the hedge-sparrows, etc., can be induced to sit at all on the egg of the cuckoo without being scandalized at the vast disproportioned size of the supposititious egg;[108] but the brute creation, I suppose, have very little idea of size, colour, or number. For the common hen, I know, when the fury of incubation is on her, will sit on a single shapeless stone instead of a nest full of eggs that have been withdrawn: and, moreover, a hen-turkey, in the same circum-stances, would sit on in the empty nest till she perished with hunger.

I think the matter might easily be determined whether a

cuckoo lays one or two eggs, or more, in a season, by opening a female during the laying-time. If more than one was come down out of the ovary, and advanced to a good size, doubtless then she would that spring lay more than one.

I will endeavour to get a hen, and to examine.

Your supposition that there may be some natural obstruction in singing birds while they are mute, and that when this is removed the song recommences, is new and bold; I wish you could discover some good grounds for this suspicion.

I was glad you were pleased with my specimen of the *caprimulgus*, or fern-owl; you were, I find, acquainted with the bird before.

When we meet, I shall be glad to have some conversation with you concerning the proposal you make of my drawing up an account of the animals in this neighbourhood.[109] Your partiality towards my small abilities persuades you, I fear, that I am able to do more than is in my power: for it is no small undertaking for a man unsupported and alone to begin a natural history from his own autopsia! Though there is endless room for observation in the field of nature, which is boundless, yet investigation (where a man endeavours to be sure of his facts) can make but slow progress; and all that one could collect in many years would go into a very narrow compass.

Some extracts from your ingenious 'Investigations of the difference between the present temperature of the air in Italy,' etc., have fallen in my way; and gave me great satisfaction: they have removed the objections that always rose in my mind whenever I came to the passages which you quote. Surely the judicious Virgil, when writing a didactic poem for the region of Italy, could never think of describing freezing rivers, unless such severity of weather pretty frequently occurred!

P.S. Swallows appear amidst snows and frost.

LETTER VI

Selborne, May 21, 1770.

DEAR SIR,

The severity and turbulence of last month so interrupted the regular progress of summer migration, that some of the birds do but just begin to show themselves, and others are apparently thinner than usual; as the white-throat, the black-cap, the red-start, the fly-catcher. I well remember that after the very severe spring in the year 1739–40 summer birds of passage were very scarce. They come probably hither with a south-east wind, or when it blows between those points; but in that unfavourable year the winds blowed the whole spring and summer through from the opposite quarters. And yet amidst all these disadvantages two swallows, as I mentioned in my last, appeared this year as early as the eleventh of April amidst frost and snow; but they withdrew again for a time.

I am not pleased to find that some people seem so little satisfied with Scopoli's new publication;* there is room to expect great things from the hands of that man, who is a good naturalist: and one would think that an history of the birds of so distant and southern a region as Carniola would be new and interesting. I could wish to see that work, and hope to get it sent down. Dr Scopoli is physician to the wretches that work in the quicksilver mines of that district.

When you talked of keeping a reed-sparrow, and giving it seeds, I could not help wondering; because the reed-sparrow which I mentioned to you (*passer arundinaceus minor Raii*) is a soft-billed bird; and most probably migrates hence before winter; whereas the bird you kept (*passer torquatus Raii*) abides all the year, and is a thick-billed bird. I question whether the latter be much of a songster; but in this matter I want to be better informed. The former has a variety of hurrying notes,

* This work he calls his *Annus Primus Historico-Naturalis.*

and sings all night. Some part of the song of the former, I suspect, is attributed to the latter. We have plenty of the soft-billed sort; which Mr Pennant had entirely left out of his *British Zoology*, till I reminded him of his omission. See *British Zoology* last published, p. 16.*

I have somewhat to advance on the different manners in which different birds fly and walk; but as this is a subject that I have not enough considered, and is of such a nature as not to be contained in a small space, I shall say nothing farther about it at present.†

No doubt the reason why the sex of birds in their first plumage is so difficult to be distinguished is, as you say, 'because they are not to pair and discharge their parental functions till the ensuing spring.' As colours seem to be the chief external sexual distinction in many birds, these colours do not take place till sexual attachments begin to obtain. And the case is the same in quadrupeds; among whom, in their younger days, the sexes differ but little: but, as they advance to maturity, horns and shaggy manes, beards and brawny necks, etc., etc., strongly discriminate the male from the female. We may instance still farther in our own species, where a beard and stronger features are usually characteristic of the male sex: but this sexual diversity does not take place in earlier life; for a beautiful youth shall be so like a beautiful girl that the difference shall not be discernible:

> Quem si puellarum insereres choro,
> Mirè sagaces falleret hospites
> Discrimen obscurum, solutis
> Crinibus, ambiguoque vultu.
>
> HOR.

---

*See Letter XXV to Mr Pennant.
†See Letter XLIII to Mr Barrington.

LETTER VII

Ringmer, near Lewes, Oct. 8, 1770.

DEAR SIR,

I am glad to hear that Kuckalm is to furnish you with the birds of Jamaica; a sight of the *hirundines* of that hot and distant island would be great entertainment to me.

The *Anni* of Scopoli are now in my possession; and I have read the *Annus Primus* with satisfaction: for though some parts of this work are exceptionable, and he may advance some mistaken observations; yet the ornithology of so distant a country as Carniola is very curious. Men that undertake only one district are much more likely to advance natural knowledge than those that grasp at more than they can possibly be acquainted with: every kingdom, every province, should have its own monographer.

The reason perhaps why he mentions nothing of Ray's Ornithology may be the extreme poverty and distance of his country, into which the works of our great naturalist may have never yet found their way. You have doubts, I know, whether this Ornithology is genuine, and really the work of Scopoli: as to myself, I think I discover strong tokens of authenticity; the style corresponds with that of his Entomology: and his characters of his Ordines and Genera are many of them new, expressive, and masterly. He has ventured to alter some of the Linnæ?n genera with sufficient show of reason.

It might perhaps be mere accident that you saw so many swifts and no swallows at Staines; because, in my long observations of those birds, I never could discover the least degree of rivalry or hostility between the species.

Ray remarks that birds of the *gallinæ* order, as cocks and hens, partridges, and pheasants, etc., are pulveratrices, such as dust themselves, using that method of cleansing their feathers, and ridding themselves of their vermin. As far as I can observe, many birds that dust themselves never wash: and I once thought that those birds that wash themselves would never dust; but here I find myself mistaken; for common house-sparrows are great pulveratrices, being frequently seen grovelling and wallowing in dusty roads; and yet they are great washers. Does not the skylark dust?

Query. – Might not Mahomet and his followers take one method of purification from these pulveratrices? because I find from travellers of credit, that if a strict Mussulman is journeying in a sandy desert where no water is to be found, at stated hours he strips off his clothes, and most scrupulously rubs his body over with sand or dust.

A countryman told me he had found a young fern-owl in the nest of a small bird on the ground; and that it was fed by the little bird. I went to see this extraordinary phenomenon, and found that it was a young cuckoo hatched in the nest of a titlark; it was become vastly too big for its nest, appearing

> . . . in tenui re
> Majores pennas nido extendisse . . .

and was very fierce and pugnacious, pursuing my finger, as I teased it, for many feet from the nest, and sparring and buffeting with its wings like a game-cock. The dupe of a dam appeared at a distance, hovering about with meat in its mouth, and expressing the greatest solicitude.

In July I saw several cuckoos skimming over a large pond; and found, after some observation, that they were feeding on the *libellulæ*, or dragon-flies; some of which they caught as they settled on the weeds, and some as they were on the wing. Notwithstanding what Linnæus says, I cannot be induced to believe that they are birds of prey.[110]

This district affords some birds that are hardly ever heard of at Selborne. In the first place considerable flocks of cross-beaks (*loxiæ curvirostræ*) have appeared this summer in the pine-groves belonging to this house; the water-ousel[111] is said to haunt the mouth of the Lewes river, near Newhaven; and the Cornish chough builds, I know, all along the chalky cliffs of the Sussex shore.

I was greatly pleased to see little parties of ring-ousels (my newly-discovered migrators) scattered, at intervals, all along the Sussex-downs from Chichester to Lewes. Let them come from whence they will, it looks very suspicious that they are cantoned along the coast in order to pass the channel when severe weather advances. They visit us again in April, as it should seem, in their return; and are not to be found in the dead of winter. It is remarkable that they are very tame, and seem to have no manner of apprehensions of danger from a person with a gun. There are bustards on the wide downs near Brighthelmstone.[112] No doubt you are acquainted with the Sussex-downs: the prospects and rides round Lewes are most lovely!

As I rode along near the coast I kept a very sharp look-out in the lanes and woods, hoping I might, at this time of the year, have discovered some of the summer short-winged birds of passage crowding towards the coast in order for their departure: but it was very extraordinary that I never saw a red-start, white-throat, black-cap, uncrested wren, fly-catcher, etc. And I remember to have made the same remark in former years, as I usually come to this place annually about this time. The birds most common along the coast at present are the stone-chatters, whin-chats, buntings, linnets, some few wheatears, titlarks, etc. Swallows and house-martins abound yet, induced to prolong their stay by this soft, still, dry season.

A land-tortoise,[113] which has been kept for thirty years in a little walled court belonging to the house where I now am visiting, retires under ground about the middle of November, and comes forth again about the middle of April. When it first appears in the spring it discovers very little inclination towards food; but in the height of summer grows voracious: and then as the summer declines its appetite declines; so that for the last six weeks in autumn it hardly eats at all. Milky plants, such as lettuces, dandelions, sow-thistles, are its favourite dish. In a neighbouring village one was kept till by tradition it was supposed to be an hundred years old. An instance of vast longevity in such a poor reptile!

LETTER VIII

Selborne, Dec. 20, 1770.

DEAR SIR,

The birds that I took for *aberdavines*[114] were reed-sparrows (*passeres torquati*).

There are doubtless many home internal migrations within this kingdom that want to be better understood: witness those vast flocks of hen chaffinches that appear with us in the winter without hardly any cocks among them. Now was there a due proportion of each sex, it should seem very improbable that any one district should produce such numbers of these little birds; and much more when only half of the species appears: therefore we may conclude that the *fringillæ cælebes*, for some good purposes, have a peculiar migration of their own in which the sexes part. Nor should it seem so wonderful that the intercourse of sexes in this species of birds should be interrupted in winter; since in many animals, and particularly in bucks and does, the sexes herd separately, except at the season when commerce is necessary for the continuance of the breed. For this matter of the chaffinches see *Fauna Suecica*, p. 85, and *Systema Naturæ*, p. 318. I see every winter vast flights of hen chaffinches, but none of cocks.

Your method of accounting for the periodical motions of the British singing birds, or birds of flight, is a very probable one; since the matter of food is a great regulator of the actions and proceedings of the brute creation: there is but one that can be set in competition with it, and that is love. But I cannot quite acquiesce with you in one circumstance when you advance that, 'when they have thus feasted, they again separate into small parties of five or six, and get the best fare they can within a certain district, having no inducement to go in quest of fresh-turned earth.' Now if you mean that the business of congregating is quite at an end from the conclusion of wheat-sowing to the season of barley and oats, it is not the case with us; for larks and chaffinches, and particularly linnets, flock and congregate as much in the very dead of winter as when the husbandman is busy with his ploughs and harrows.

Sure there can be no doubt but that woodcocks and fieldfares leave us in the spring, in order to cross the seas, and to retire to some districts more suitable to the purpose of breeding. That the former pair before they retire, and that the hens are forward with egg, I myself, when I was a sportsman, have often experienced. It cannot indeed be denied but that now and then we hear of a woodcock's nest,[115] or young birds, discovered in some part or other of this island: but then they are always mentioned as rarities, and somewhat out of the common course of things: but as to redwings[116] and fieldfares, no sportsman or naturalist has ever yet, that I could hear, pretended to have found the nest or young of those species in any part of these kingdoms. And I the more admire at this instance as extraordinary, since, to all appearance, the same food in summer as well as in winter might support them here which maintains their congeners, the blackbirds and thrushes, did they choose to stay the summer through. From hence it appears that it is not food alone which determines some species of birds with regard to their stay or departure. Fieldfares and red-wings disappear sooner or later according as the warm weather comes on earlier or later. For I well remember, after that dreadful winter of 1739–40, that cold north-east winds continued to blow on through April and May, and that these kinds of birds (what few remained of them) did not depart as usual, but were seen lingering about till the beginning of June.

The best authority that we can have for the nidification of the birds above-mentioned in any district, is the testimony of faunists that have written professedly the natural history of particular countries. Now, as to the fieldfare, Linnæus, in his *Fauna Suecica*, says of it that '*maximis in arboribus nidificat*';[117] and of the redwing he says, in the same place, that '*nidificat in mediis arbusculis, sive sepibus: ova sex cæruleo-viridia maculis nigris variis.*' Hence we may be assured that fieldfares and redwings breed in Sweden. Scopoli says, in his *Annus Primus*, of the woodcock, that '*nupta ad nos venit circa æquinoctium vernale*'; meaning in Tirol, of which he is a native. And afterwards he adds '*nidificat in paludibus alpinis: ova ponit*, 3–5.' It does not appear from Kramer that woodcocks breed at all in Austria: but he says '*Avis hæc septentrionalium provinciarum æstivo tempore incola est; ubi plerumque nidificat. Appropinquante hyeme australiores provincias petit: hinc circa plenilunium mensis Octobris plerumque Austriam transmigrat. Tunc rursus circa plenilunium potissimum mensis Martii per Austriam matrimonio juncta ad septentrionales provincias redit.*' For the whole passage (which I have abridged) see *Elenchus*, etc., p. 351. This seems to be a full proof of the migration of woodcocks; though little is proved concerning the place of breeding.

P.S. There fell in the county of Rutland, in three weeks of this present very wet weather, seven inches and an half of rain, which is more than has fallen in any three weeks for these thirty years past in that part of the world. A mean quantity in that county for one year is twenty inches and an half.

LETTER IX

Fyfield, near Andover, Feb. 12, 1771.

DEAR SIR,

You are, I know, no great friend to migration; and the well attested accounts from various parts of the kingdom seem to justify you in your suspicions, that at least many of the swallow kind do not leave us in the winter, but lay themselves up like insects and bats, in a torpid state, to slumber away the more uncomfortable months till the return of the sun and fine weather awakens them.

But then we must not, I think, deny migration in general; because migration certainly does subsist in some places, as my brother in Andalusia has fully informed me. Of the motions of these birds he has ocular demonstration for many weeks together, both spring and fall: during which periods myriads of the swallow kind traverse the Straits from north to south, and from south to north, according to the season. And these vast migrations consist not only of *hirundines* but of bee-birds, hoopoes, *oro pendolos* or golden thrushes,[118] etc., etc., and also many of our soft-billed summer-birds of passage; and moreover of birds which never leave us, such as all the various sorts of hawks and kites. Old Belon, two hundred years ago, gives a curious account of the incredible armies of hawks and kites which he saw in the spring-time traversing the Thracian Bosphorus from Asia to Europe. Besides the above-mentioned, he remarks that the procession is swelled by whole troops of eagles and vultures.

Now it is no wonder that birds residing in Africa should retreat before the sun as it advances, and retire to milder regions, and especially birds of prey, whose blood being heated with hot animal food, are more impatient of a sultry climate: but then I cannot help wondering why kites and hawks, and such hardy birds as are known to defy all the severity of England,

and even of Sweden and all north Europe, should want to migrate from the south of Europe, and be dissatisfied with the winters of Andalusia.

It does not appear to me that much stress may be laid on the difficulty and hazard that birds must run in their migrations, by reason of vast oceans, cross winds, etc.; because, if we reflect, a bird may travel from England to the equator without launching out and exposing itself to boundless seas, and that by crossing the water at Dover, and again at Gibraltar. And I with the more confidence advance this obvious remark, because my brother has always found that some of his birds, and particularly the swallow kind, are very sparing of their pains in crossing the Mediterranean: for when arrived at Gibraltar, they do not

> ... rang'd in figure wedge their way,
> ... and set forth
> Their airy caravan high over seas
> Flying, and over lands with mutual wing
> Easing their flight ...
>
> MILTON.

but scout and hurry along in little detached parties of six or seven in a company; and sweeping low, just over the surface of the land and water, direct their course to the opposite continent at the narrowest passage they can find. They usually slope across the bay to the south-west, and so pass over opposite to Tangier, which, it seems, is the narrowest space.

In former letters we have considered whether it was probable that woodcocks in moon-shiny nights cross the German ocean from Scandinavia. As a proof that birds of less speed may pass that sea, considerable as it is, I shall relate the following incident, which, though mentioned to have happened so many years ago, was strictly matter of fact: As some people were shooting in the parish of Trotton, in the county of Sussex, they killed a duck in that dreadful winter 1708–9, with a silver collar about its neck,*

---

*I have read a like anecdote of a swan.

on which were engraven the arms of the king of Denmark. This anecdote the rector of Trotton at that time has often told to a near relation of mine; and, to the best of my remembrance, the collar was in the possession of the rector.

At present I do not know anybody near the sea-side that will take the trouble to remark at what time of the moon wood-cocks first come: if I lived near the sea myself I would soon tell you more of the matter. One thing I used to observe when I was a sportsman, that there were times in which woodcocks were so sluggish and sleepy that they would drop again when flushed just before the spaniels, nay, just at the muzzle of a gun that had been fired at them: whether this strange laziness was the effect of a recent fatiguing journey I shall not presume to say.

Nightingales not only never reach Northumberland and Scotland, but also, as I have been always told, Devonshire and Cornwall. In those two last counties we cannot attribute the failure of them to the want of warmth: the defect in the west is rather a presumptive argument that these birds come over to us from the continent at the narrowest passage, and do not stroll so far westward.

Let me hear from your own observation whether skylarks do not dust. I think they do: and if they do, whether they wash also.

The *alauda pratensis*[119] of Ray was the poor dupe that was educating the booby of a cuckoo mentioned in my letter of October last.

Your letter came too late for me to procure a ring-ousel for Mr Tunstal during their autumnal visit; but I will endeavour to get him one when they call on us again in April. I am glad that you and that gentleman saw my Andalusian birds; I hope they answered your expectation. Royston, or grey crows, are winter birds that come much about the same time with the woodcock: they, like the fieldfare and redwing, have no apparent reason for migration; for as they fare in the winter like their congeners, so might they in all appearance in the summer. Was not Tenant, when a boy, mistaken? did he not find a missel-thrush's nest, and take it for the nest of a fieldfare?

The stock-dove, or wood-pigeon, *œnas Raii*, is the last winter bird of passage which appears with us; and is not seen till towards the end of November: about twenty years ago they abounded in the district of Selborne; and strings of them were seen morning and evening that reached a mile or more: but since the beechen woods have been greatly thinned they are much decreased in number. The ring-dove, *palumbus Raii*, stays with us the whole year, and breeds several times through the summer.

Before I received your letter of October last I had just remarked in my journal that the trees were unusually green. This uncommon verdure lasted on late into November; and may be accounted for from a late spring, a cool and moist summer; but more particularly from vast armies of chafers, or tree beetles, which, in many places, reduced whole woods to a leafless naked state. These trees shot again at Midsummer, and then retained their foliage till very late in the year.

My musical friend,[120] at whose house I am now visiting, has tried all the owls that are his near neighbours with a pitch-pipe, set at concert-pitch, and finds they all hoot in B flat. He will examine the nightingales next spring.

I am, etc., etc.

LETTER X

Selborne, Aug. 1, 1771.

DEAR SIR,

From what follows, it will appear that neither owls nor cuckoos keep to one note. A friend remarks that many (most) of his owls hoot in B flat: but that one went almost half a note below A. The pipe he tried their notes by was a common half-crown pitch-pipe, such as masters use for tuning of harpsichords; it was the common London pitch.

A neighbour of mine, who is said to have a nice ear, remarks that the owls about this village hoot in three different keys, in G flat, or F sharp, in B flat and A flat. He heard two hooting to each other, the one in A flat, and the other in B flat. *Query:* Do these different notes proceed from different species, or only from various individuals? The same person finds upon trial that the note of the cuckoo (of which we have but one species) varies in different individuals; for, about Selborne wood, he found they were mostly in D: he heard two sing together, the one in D, the other in D sharp, who made a disagreeable concert: he afterwards heard one in D sharp, and about Wolmer-forest some in C. As to nightingales, he says that their notes are so short, and their transitions so rapid, that he cannot well ascertain their key. Perhaps in a cage, and in a room, their notes may be more distinguishable. This person has tried to settle the notes of a swift, and of several other small birds, but cannot bring them to any criterion.

As I have often remarked that redwings are some of the first birds that suffer with us in severe weather, it is no wonder at all that they retreat from Scandinavian winters: and much more the *ordo* of *grallæ*, who, all to a bird, forsake the northern parts of Europe at the approach of winter. '*Grallæ tanquam conjuratæ unanimiter in fugam se conjiciunt; ne earum unicam quidem inter nos habitantem invenire possimus; ut enim æstate in australibus degere nequeunt ob defectum lumbricorum, terramque siccam; ita nec in frigidis ob eandem causam,*'[121] says Eckmarck the Swede, in his ingenious little treatise called *Migrationes Avium*, which by all means you ought to read while your thoughts run on the subject of migration. See *Amœnitates Academicæ*, vol. iv, p. 565.

Birds may be so circumstanced as to be obliged to migrate in one country and not in another: but the *grallæ* (which procure their food from marshes and boggy grounds) must in winter forsake the more northerly parts of Europe, or perish for want of food.

I am glad you are making inquiries from Linnæus concerning the woodcock: it is expected of him that he should be able to account for the motions and manner of life of the animals of his own *Fauna*.

Faunists, as you observe, are too apt to acquiesce in bare descriptions, and a few synonyms: the reason is plain; because all that may be done at home in a man's study, but the investigation of the life and conversation of animals, is a concern of much more trouble and difficulty, and is not to be attained but by the active and inquisitive, and by those that reside much in the country.

Foreign systematics are, I observe, much too vague in their specific differences; which are almost universally constituted by one or two particular marks, the rest of the description running in general terms. But our countryman, the excellent Mr Ray, is the only describer that conveys some precise idea in every term or word, maintaining his superiority over his followers and imitators in spite of the advantage of fresh discoveries and modern information.

At this distance of years it is not in my power to recollect at what periods woodcocks used to be sluggish or alert when I was a sportsman; but, upon my mentioning this circumstance to a friend, he thinks he has observed them to be remarkably listless against snowy foul weather: if this should be the case, then the inaptitude for flying arises only from an eagerness for food; as sheep are observed to be very intent on grazing against stormy wet evenings.

I am, etc., etc.

LETTER XI

Selborne, Feb. 8, 1772.

DEAR SIR,

When I ride about in the winter, and see such prodigious flocks of various kinds of birds, I cannot help admiring at these congregations, and wishing that it was in my power to account for those appearances almost peculiar to the season. The two great motives which regulate the proceedings of the brute creation are love and hunger; the former incites animals to perpetuate their kind, the latter induces them to preserve individuals; whether either of these should seem to be the ruling passion in the matter of congregating is to be considered. As to love, that is out of the question at a time of the year when that soft passion is not indulged; besides, during the amorous season, such a jealousy prevails between the male birds that they can hardly bear to be together in the same hedge or field. Most of the singing and elation of spirits of that time seem to me to be the effect of rivalry and emulation: and it is to this spirit of jealousy that I chiefly attribute the equal dispersion of birds in the spring over the face of the country.

Now as to the business of food: as these animals are actuated by instinct to hunt for necessary food, they should not, one would suppose, crowd together in pursuit of sustenance at a time when it is most likely to fail: yet such associations do take place in hard weather chiefly, and thicken as the severity increases. As some kind of self-interest and self-defence is no doubt the motive for the proceeding, may it not arise from the helplessness of their state in such rigorous seasons; as men crowd together, when under great calamities, though they know not why? Perhaps approximation may dispel some degree of cold; and a crowd may make each individual appear safer from the ravages of birds of prey and other dangers.

If I admire when I see how much congenerous birds love to

congregate, I am the more struck when I see incongruous ones in such strict amity. If we do not much wonder to see a flock of rooks usually attended by a train of daws, yet it is strange that the former should so frequently have a flight of starlings for their satellites. Is it because rooks have a more discerning scent than their attendants, and can lead them to spots more productive of food? Anatomists say that rooks, by reason of two large nerves which run down between the eyes into the upper mandible, have a more delicate feeling in their beaks than other round-billed birds, and can grope for their meat when out of sight. Perhaps then their associates attend them on the motive of interest, as greyhounds wait on the motions of their finders; and as lions are said to do on the yelpings of jackals. Lapwings and starlings sometimes associate.

LETTER XII

March 9, 1772.

DEAR SIR,

As a gentleman and myself were walking on the fourth of last November round the sea-banks at Newhaven, near the mouth of the Lewes river, in pursuit of natural knowledge, we were surprised to see three house-swallows gliding very swiftly by us. That morning was rather chilly, with the wind at northwest; but the tenor of the weather for some time before had been delicate, and the noons remarkably warm. From this incident, and from repeated accounts which I meet with, I am more and more induced to believe that many of the swallow kind do not depart from this island; but lay themselves up in holes and caverns; and do, insect-like and bat-like, come forth at mild times, and then retire again to their *latebræ*. Nor make I the least doubt but that, if I lived at Newhaven, Seaford, Brighthelmstone, or any of those towns near the chalk-cliffs of the

Sussex coast, by proper observations, I should see swallows stirring at periods of the winter, when the noons were soft and inviting, and the sun warm and invigorating. And I am the more of this opinion from what I have remarked during some of our late springs, that though some swallows did make their appearance about the usual time, viz., the thirteenth or fourteenth of April, yet meeting with an harsh reception, and blustering cold north-east winds, they immediately withdrew, absconding for several days, till the weather gave them better encouragement.

LETTER XIII

April 12, 1772.

DEAR SIR,

While I was in Sussex last autumn my residence was at the village near Lewes, from whence I had formerly the pleasure of writing to you. On the first of November I remarked that the old tortoise, formerly mentioned, began first to dig the ground in order to the forming its hybernaculum, which it had fixed on just beside a great tuft of hepaticas. It scrapes out the ground with its fore-feet, and throws it up over its back with its hind; but the motion of its legs is ridiculously slow, little exceeding the hour-hand of a clock; and suitable to the composure of an animal said to be a whole month in performing one feat of copulation. Nothing can be more assiduous than this creature night and day in scooping the earth, and forcing its great body into the cavity; but, as the noons of that season proved unusually warm and sunny, it was continually interrupted, and called forth by the heat in the middle of the day; and though I continued there till the thirteenth of November, yet the work remained unfinished. Harsher weather, and frosty mornings, would have quickened its operations. No part of its behaviour ever struck me more than the extreme timidity it always expresses with regard to rain; for though it has a shell that would secure it against the wheel of a loaded cart, yet does it discover as much solicitude about rain as a lady dressed in all her best attire, shuffling away on the first sprinklings, and running its head up in a corner. If attended to, it becomes an excellent weather-glass; for as sure as it walks elate, and as it were on tiptoe, feeding with great earnestness in a morning, so sure will it rain before night. It is totally a diurnal animal, and never pretends to stir after it becomes dark. The tortoise, like other reptiles, has an arbitrary stomach as well as lungs; and can refrain from eating as well as breathing for a great part of the year. When first awakened it eats nothing; nor again in the autumn before it retires: through the height of the summer it feeds voraciously, devouring all the food that comes in its way. I was much taken with its sagacity in discerning those that do it kind offices; for, as soon as the good old lady comes in sight who has waited on it for more than thirty years, it hobbles towards its benefactress with awkward alacrity; but remains inattentive to strangers. Thus not only 'the ox knoweth his owner, and the ass his master's crib,'* but the most abject reptile and torpid of beings distinguishes the hand that feeds it, and is touched with the feelings of gratitude!

I am, etc., etc.

P.S. In about three days after I left Sussex the tortoise retired into the ground under the hepatica.

*Isaiah i. 3.

LETTER XIV

Selborne, March 26, 1773.

DEAR SIR,

The more I reflect on the στοργὴ of animals, the more I am astonished at its effects. Nor is the violence of this affection more wonderful than the shortness of its duration. Thus every hen is in her turn the virago of the yard, in proportion to the helplessness of her brood; and will fly in the face of a dog or a sow in defence of those chickens, which in a few weeks she will drive before her with relentless cruelty.

This affection sublimes the passions, quickens the invention, and sharpens the sagacity of the brute creation. Thus an hen, just become a mother, is no longer that placid bird she used to be, but with feathers standing on end, wings hovering, and clocking note, she runs about like one possessed. Dams will throw themselves in the way of the greatest danger in order to avert it from their progeny. Thus a partridge will tumble along before a sportsman in order to draw away the dogs from her helpless covey. In the time of nidification the most feeble birds will assault the most rapacious. All the *hirundines* of a village are up in arms at the sight of an hawk, whom they will persecute till he leaves that district. A very exact observer has often remarked that a pair of ravens nesting in the rock of Gibraltar would suffer no vulture or eagle to rest near their station, but would drive them from the hill with an amazing fury: even the blue thrush[122] at the season of breeding would dart out from the clefts of the rocks to chase away the kestril, or the sparrow-hawk. If you stand near the nest of a bird that has young, she will not be induced to betray them by an inadvertent fondness, but will wait about at a distance with meat in her mouth for an hour together.

Should I farther corroborate what I have advanced above by some anecdotes which I probably may have mentioned before in conversation, yet you will, I trust, pardon the repetition for the sake of illustration.

The fly-catcher of the *Zoology* (the *stoparola* of Ray), builds every year in the vines that grow on the walls of my house. A pair of these little birds had one year inadvertently placed their nest on a naked bough, perhaps in a shady time; not being aware of the inconvenience that followed. But an hot sunny season coming on before the brood was half fledged, the reflection of the wall became insupportable, and must inevitably have destroyed the tender young, had not affection suggested an expedient, and prompted the parent-birds to hover over the nest all the hotter hours, while with wings expanded, and mouths gaping for breath, they screened off the heat from their suffering offspring.

A farther instance I once saw of notable sagacity in a willow-wren, which had built in a bank in my fields. This bird a friend and myself had observed as she sat in her nest; but were particularly careful not to disturb her, though we saw she eyed us with some degree of jealousy. Some days after as we passed that way we were desirous of remarking how this brood went on; but no nest could be found, till I happened to take up a large bundle of long green moss, as it were, carelessly thrown over the nest, in order to dodge the eye of any impertinent intruder.

A still more remarkable mixture of sagacity and instinct occurred to me one day as my people were pulling off the lining of an hotbed, in order to add some fresh dung. From out of the side of this bed leaped an animal with great agility that made a most grotesque figure; nor was it without great difficulty that it could be taken; when it proved to be a large white-bellied field-mouse[123] with three or four young clinging to her teats by their mouths and feet. It was amazing that the desultory and rapid motions of this dam should not oblige her litter to quit their hold, especially when it appeared that they were so young as to be both naked and blind!

To these instances of tender attachment, many more of which might be daily discovered by those that are studious of nature, may be opposed that rage of affection, that monstrous perversion of the στοργή, which induces some females of the brute creation to devour their young because their owners have handled them too freely, or removed them from place to place! Swine, and sometimes the more gentle race of dogs and cats, are guilty of this horrid and preposterous murder. When I hear now and then of an abandoned mother that destroys her offspring, I am not so much amazed; since reason perverted, and the bad passions let loose, are capable of any enormity: but why the parental feelings of brutes, that usually flow in one most uniform tenor, should sometimes be so extravagantly diverted, I leave to abler philosophers than myself to determine.

I am, etc.

LETTER XV

Selborne, July 8, 1773.

DEAR SIR,

Some young men went down lately to a pond on the verge of Wolmer-forest to hunt flappers, or young wild-ducks, many of which they caught, and, among the rest, some very minute yet well-fledged wild-fowls alive, which, upon examination, I found to be teals. I did not know till then that teals ever bred in the south of England, and was much pleased with the discovery: this I look upon as a great stroke in natural history.

We have had, ever since I can remember, a pair of white owls that constantly breed under the eaves of this church. As I have paid good attention to the manner of life of these birds during their season of breeding, which lasts the summer through, the following remarks may not perhaps be unacceptable: About an hour before sunset (for then the mice begin to run) they sally forth in quest of prey, and hunt all round the hedges of meadows and small enclosures for them, which seem to be their only food. In this irregular country we can stand on an eminence and see them beat the fields over like a setting-dog, and often drop down in the grass or corn. I have minuted these birds with my watch for an hour together, and have found that they return to their nests, the one or the other of them, about once in five minutes; reflecting at the same time on the adroitness that every animal is possessed of as regards the well-being of itself and off-spring. But a piece of address, which they show when they return loaded, should not, I think, be passed over in silence. – As they take their prey with their claws, so they carry it in their claws to their nest: but, as the feet are necessary in their ascent under the tiles, they constantly perch first on the roof of the chancel, and shift the mouse from their claws to their bill, that the feet may be at liberty to take hold of the plate on the wall as they are rising under the eaves.

White owls seem not (but in this I am not positive) to hoot at all: all that clamorous hooting appears to me to come from the wood kinds. The white owl does indeed snore and hiss in a tremendous manner; and these menaces will answer the intention of intimidating: for I have known a whole village up in arms on such an occasion, imagining the church-yard to be full of goblins and spectres. White owls also often scream horribly as they fly along; from this screaming probably arose the common people's imaginary species of screech-owl, which they superstitiously think attends the windows of dying persons. The plumage of the remiges of the wings of every species of owl that I have yet examined is remarkably soft and pliant. Perhaps it may be necessary that the wings of these birds should not make much resistance or rushing, that they may be enabled to steal through the air unheard upon a nimble and watchful quarry.

While I am talking of owls, it may not be improper to mention what I was told by a gentleman of the county of Wilts. As they were grubbing a vast hollow pollard-ash that had been the mansion of owls for centuries, he discovered at the bottom a mass of matter that at first he could not account for. After some examination, he found it was a congeries of the bones of mice (and perhaps of birds and bats) that had been heaping together for ages, being cast up in pellets out of the crops of many generations of inhabitants. For owls cast up the bones, fur and feathers of what they devour, after the manner of hawks. He believes, he told me, that there were bushels of this kind of substance.

When brown owls hoot their throats swell as big as an hen's egg. I have known an owl of this species live a full year without any water. Perhaps the case may be the same with all birds of prey. When owls fly they stretch out their legs behind them as a balance to their large heavy heads; for as most nocturnal birds have large eyes and ears they must have large heads to contain them. Large eyes I presume are necessary to collect every ray of light, and large concave ears to command the smallest degree of sound or noise.

I am, etc.

———————

It will be proper to premise here that the sixteenth, eighteenth, twentieth, and twenty-first letters have been published already in the *Philosophical Transactions*: but as nicer observation has furnished several corrections and additions, it is hoped that the republication of them will not give offence; especially as these sheets would be very imperfect without them, and as they will be new to many readers who had no opportunity of seeing them when they made their first appearance.

The *hirundines* are a most inoffensive, harmless, entertaining, social, and useful tribe of birds: they touch no fruit in our gardens; delight, all except one species, in attaching themselves to our houses; amuse us with their migrations, songs, and marvellous agility; and clear our outlets from the annoyance of gnats and other troublesome insects. Some districts in the south seas, near Guiaquil,* are desolated, it seems, by the infinite swarms of venomous mosquitoes, which fill the air, and render those coasts insupportable. It would be worth inquiring whether any species of *hirundines* is found in those regions. Whoever contemplates the myriads of insects that sport in the sunbeams of a summer evening in this country, will soon be convinced to what a degree our atmosphere would be choked with them was it not for the friendly interposition of the swallow tribe.

Many species of birds have their particular lice; but the *hirundines* alone seem to be annoyed with dipterous insects, which infest every species, and are so large, in proportion to themselves, that they must be truly irksome and injurious to them. These are the *hippoboscæ hirundinis*[124] with narrow subulated wings, abounding in every nest; and are hatched by the warmth of the bird's own body during incubation, and crawl about under its feathers.

A species of them is familiar to horsemen in the south of England under the name of forest-fly; and, to some, of side-fly, from its running sideways like a crab. It creeps under the tails, and about the groins, of horses, which, at their first coming out of the north, are rendered half frantic by the tickling sensation; while our own breed little regards them.

The curious Réaumur[125] discovered the large eggs, or rather *pupæ*, of these flies as big as the flies themselves, which he hatched in his own bosom. Any person that will take the trouble to examine the old nests of either species of swallows may find in them the black shining cases of the *pupæ* of these insects: but for other particulars, too long for this place, we refer the reader to *L'Histoire d'Insectes* of that admirable entomologist. Tom. iv. pl. ii.

---

*See Ulloa's *Travels*.

LETTER XVI

Selborne, Nov. 20, 1773.

DEAR SIR,

In obedience to your injunctions I sit down to give you some account of the house-martin, or martlet; and, if my monography of this little domestic and familiar bird should happen to meet with your approbation, I may probably soon extend my inquiries to the rest of the British *hirundines* – the swallow, the swift, and the bank-martin.

A few house-martins begin to appear about the sixteenth of April; usually some few days later than the swallow. For some time after they appear the *hirundines* in general pay no attention to the business of nidification, but play and sport about either to recruit from the fatigue of their journey, if they do migrate at all, or else that their blood may recover its true tone and texture after it has been so long benumbed by the severities of winter. About the middle of May, if the weather be fine, the martin begins to think in earnest of providing a mansion for its family. The crust or shell of this nest seems to be formed of such dirt or loam as comes most readily to hand, and is tempered and wrought together with little bits of broken straws to render it tough and tenacious. As this bird often builds against a perpendicular wall without any projecting ledge under, it requires its utmost efforts to get the first foundation firmly fixed, so that it may safely carry the superstructure. On this occasion the bird not only clings with its claws, but partly supports itself by strongly inclining its tail against the wall, making that a fulcrum; and thus steadied it works and plasters the materials into the face of the brick or stone. But then, that this work may not, while it is soft and green, pull itself down by its own weight, the provident architect has prudence and forbearance enough not to advance her work too fast; but by building only in the morning, and by dedicating the rest of the day to food and amusement, gives it sufficient time to dry and harden. About half an inch seems to be a sufficient layer for a day. Thus careful workmen when they build mud-walls (informed at first perhaps by this little bird) raise but a moderate layer at a time, and then desist; lest the work should become top-heavy, and so be ruined by its own ·weight. By this method in about ten or twelve days is formed an hemispheric nest with a small aperture towards the top, strong, compact, and warm; and perfectly fitted for all the purposes for which it was intended. But then nothing is more common than for the house-sparrow, as soon as the shell is finished, to seize on it as its own, to eject the owner, and to line it after its own manner.

After so much labour is bestowed in erecting a mansion, as nature seldom works in vain, martins will breed on for several years together in the same nest, where it happens to be well sheltered and secure from the injuries of weather. The shell or crust of the nest is a sort of rustic work full of knobs and protuberances on the outside: nor is the inside of those that I have examined smoothed with any exactness at all; but is rendered soft and warm, and fit for incubation, by a lining of small straws, grasses, and feathers; and sometimes by a bed of moss interwoven with wool. In this nest they tread, or engender, frequently during the time of building; and the hen lays from three to five white eggs.

At first when the young are hatched, and are in a naked and helpless condition, the parent birds, with tender assiduity, carry out what comes away from their young. Was it not for this affectionate cleanliness the nestlings would soon be burnt up, and destroyed in so deep and hollow a nest, by their own caustic excrement. In the quadruped creation the same neat precaution is made use of; particularly among dogs and cats, where the dams lick away what proceeds from their young. But in birds there seems to be a particular provision, that the dung of nestlings is enveloped into a tough kind of jelly, and therefore is the easier conveyed off without soiling or daubing. Yet, as nature is cleanly in all her ways, the young perform this office for themselves in a little time by thrusting their tails out at the aperture of their nest. As the young of small birds presently arrive at their ἡλικία or full growth, they soon become impatient of confinement, and sit all day with their heads out at the orifice, where the dams, by clinging to the nest, supply them with food from morning to night. For a time the young are fed on the wing by their parents; but the feat is done by so quick and almost imperceptible a sleight, that a person must have attended very exactly to their motions before he would be able to perceive it. As soon as the young are able to shift for themselves, the dams[126] immediately turn their thoughts to the business of a second brood: while the first flight, shaken off and rejected by their nurses, congregate in great flocks, and are the birds that are seen clustering and hovering on sunny mornings and evenings round towers and steeples, and on the roofs of churches and houses. These congregations usually begin to take place about the first week in August; and therefore we may conclude that by that time the first flight is pretty well over. The young of this species do not quit their abodes all together; but the more forward birds get abroad some days before the rest. These approaching the eaves of buildings, and playing about before them, make people think that several old ones attend one nest. They are often capricious in fixing on a nesting place, beginning many edifices, and leaving them unfinished; but when once a nest is completed in a sheltered place, it serves for several seasons. Those which breed in a ready finished house get the start in hatching of those that build new by ten days or a fortnight. These industrious artificers are at their labours in the long days before four in the morning: when they fix their materials they plaster them on with their chins, moving their heads with a quick vibratory motion. They dip and wash as they fly sometimes in very hot weather, but not so frequently as swallows. It has been observed that martins usually build to a north-east or north-west aspect, that the heat of the sun may not crack and destroy their nests: but instances are also remembered where they bred for many years in vast abundance in an hot stifled inn-yard, against a wall facing to the south.

Birds in general are wise in their choice of situation: but in this neighbourhood every summer is seen a strong proof to the contrary at an house without eaves in an exposed district, where some martins build year by year in the corners of the windows. But, as the corners of these windows (which face to the south-east and south-west) are too shallow, the nests are washed down every hard rain; and yet these birds drudge on to no purpose from summer to summer, without changing their aspect or house. It is a piteous sight to see them labouring when half their nest is washed away and bringing dirt . . . 'generis lapsi sarcire ruinas.' Thus is instinct a most wonderful unequal faculty; in some instances so much above reason, in other respects so far below it! Martins love to frequent towns, especially if there are great lakes and rivers at hand; nay, they even affect the close air of London. And I have not only seen them nesting in the Borough, but even in the Strand and Fleet-street; but then it was obvious from the dinginess of their aspect that their feathers partook of the filth of that sooty atmosphere. Martins are by far the least agile of the four species; their wings and tails are short, and therefore they are not capable of such surprising turns and quick and glancing evolutions as the swallow. Accordingly they make use of a placid easy motion in a middle region of the air, seldom mounting to any great height, and never sweeping long together over the surface of the ground or water. They do not wander far for food, but affect sheltered districts, over some lake, or under some hanging wood, or in some hollow vale, especially in windy weather. They breed the latest of all the swallow kind: in 1772 they had nestlings on to October the twenty-first, and are never without unfledged young as late as Michaelmas.

As the summer declines the congregating flocks increase in numbers daily by the constant accession of the second broods;

till at last they swarm in myriads upon myriads round the villages on the Thames, darkening the face of the sky as they frequent the aits of that river, where they roost. They retire, the bulk of them I mean, in vast flocks together about the beginning of October: but have appeared of late years in a considerable flight in this neighbourhood, for one day or two, as late as November the third and sixth, after they were supposed to have been gone for more than a fortnight. They therefore withdraw with us the latest of any species. Unless these birds are very short-lived indeed, or unless they do not return to the district where they are bred, they must undergo vast devastations somehow, and somewhere; for the birds that return yearly bear no manner of proportion to the birds that retire.

House-martins are distinguished from their congeners by having their legs covered with soft downy feathers down to their toes. They are no songsters; but twitter in a pretty inward soft manner in their nests. During the time of breeding they are often greatly molested with fleas.

I am, etc.

LETTER XVII

Ringmer, near Lewes, Dec. 9, 1773.

DEAR SIR,

I received your last favour just as I was setting out for this place; and am pleased to find that my monography met with your approbation. My remarks are the result of many years' observation; and are, I trust, true on the whole: though I do not pretend to say that they are perfectly void of mistake, or that a more nice observer might not make many additions, since subjects of this kind are inexhaustible.

If you think my letter worthy the notice of your respectable society,[127] you are at liberty to lay it before them; and they will consider it, I hope, as it was intended, as an humble attempt to promote a more minute inquiry into natural history; into the life and conversation of animals. Perhaps hereafter I may be induced to take the house-swallow under consideration, and from that proceed to the rest of the British *hirundines*.

Though I have now travelled the Sussex-downs upwards of thirty years, yet I still investigate that chain of majestic mountains with fresh admiration year by year; and think I see new beauties every time I traverse it. This range, which runs from Chichester eastward as far as East-Bourn, is about sixty miles in length, and is called the South Downs, properly speaking, only round Lewes. As you pass along you command a noble view of the wild, or weald, on one hand, and the broad downs and sea on the other. Mr Ray used to visit a family* just at the foot of these hills, and was so ravished with the prospect from Plumpton-plain near Lewes, that he mentions those scapes in his *Wisdom of God in the Works of the Creation* with the utmost satisfaction, and thinks them equal to anything he had seen in the finest parts of Europe.

For my own part, I think there is somewhat peculiarly sweet and amusing in the shapely figured aspect of chalk-hills in preference to those of stone, which are rugged, broken, abrupt, and shapeless.

Perhaps I may be singular in my opinion, and not so happy as to convey to you the same idea; but I never contemplate these mountains without thinking I perceive somewhat analogous to growth in their gentle swellings and smooth fungus-like protuberances, their fluted sides, and regular hollows and slopes, that carry at once the air of vegetative dilation and expansion. . . . Or was there ever a time when these immense masses of calcareous matter were thrown into fermentation by some adventitious moisture; were raised and leavened into such shapes by some plastic power; and so made to swell and

*Mr Courthope, of Danny.

heave their broad backs into the sky so much above the less animated clay of the wild below?

By what I can guess from the admeasurements of the hills that have been taken round my house, I should suppose that these hills surmount the wild at an average at about the rate of five hundred feet.

One thing is very remarkable as to the sheep: from the westward till you get to the river Adur all the flocks have horns, and smooth white faces, and white legs; and a hornless sheep is rarely to be seen: but as soon as you pass the river eastward, and mount Beeding-hill, all the flocks at once become hornless, or, as they call them, poll-sheep; and have moreover black faces with a white tuft of wool on their foreheads, and speckled and spotted legs: so that you would think that the flocks of Laban were pasturing on one side of the stream, and the variegated breed of his son-in-law Jacob were cantoned along on the other. And this diversity holds good respectively on each side from the valley of Bramber and Beeding to the eastward, and westward all the whole length of the downs. If you talk with the shepherds on this subject, they tell you that the case has been so from time immemorial: and smile at your simplicity if you ask them whether the situation of these two different breeds might not be reversed? However, an intelligent friend of mine near Chichester is determined to try the experiment; and has this autumn, at the hazard of being laughed at, introduced a parcel of black-faced hornless rams among his horned western ewes. The black-faced poll-sheep[128] have the shortest legs and the finest wool.

As I had hardly ever before travelled these downs at so late a season of the year, I was determined to keep as sharp a look-out as possible so near the southern coast, with respect to the summer short-winged birds of passage. We make great inquiries concerning the withdrawing of the swallow kind, without examining enough into the causes why this tribe is never to be seen in winter; for, *entre nous*, the disappearing of the latter is more marvellous than that of the former, and much more unaccountable. The *hirundines*, if they please, are certainly capable of migration; and yet no doubt are often found in a torpid state: but red-starts, nightingales, white-throats, black-caps, etc., etc., are very ill provided for long flights; have never been once found, as I ever heard of, in a torpid state, and yet can never be supposed, in such troops, from year to year to dodge and elude the eyes of the curious and inquisitive, which from day to day discern the other small birds that are known to abide our winters. But, notwithstanding all my care, I saw nothing like a summer bird of passage: and, what is more strange, not one wheat-ear, though they abound so in the autumn as to be a considerable perquisite to the shepherds that take them; and though many are to be seen to my knowledge all the winter through in many parts of the south of England. The most intelligent shepherds tell me that some few of these birds appear on the downs in March, and then withdraw to breed probably in warrens and stone-quarries: now and then a nest is plowed up in a fallow on the downs under a furrow, but it is thought a rarity. At the time of wheat-harvest they begin to be taken in great numbers; are sent for sale in vast quantities to Brighthelmstone and Tunbridge; and appear at the tables of all the gentry that entertain with any degree of elegance. About Michaelmas they retire and are seen no more till March. Though these birds are, when in season, in great plenty on the south downs round Lewes, yet at East-Bourn, which is the eastern extremity of those downs, they abound much more. One thing is very remarkable – that though in the height of the season so many hundreds of dozens are taken, yet they never are seen to flock; and it is a rare thing to see more than three or four at a time: so that there must be a perpetual flitting and constant progressive succession. It does not appear that any wheat-ears are taken to the westward of Houghton-bridge, which stands on the river Arun.

I did not fail to look particularly after my new migration of ring-ousels; and to take notice whether they continued on the downs to this season of the year; as I had formerly remarked them in the month of October all the way from Chichester to Lewes wherever there were any shrubs and covert: but not one bird of this sort came within my observation. I only saw a few larks and whin-chats, some rooks, and several kites and buzzards.[129]

About Midsummer a flight of cross-bills comes to the pine-groves about this house, but never makes any long stay.

The old tortoise, that I have mentioned in a former letter, still continues in this garden; and retired under ground about the twentieth of November, and came out again for one day on the thirtieth: it lies now buried in a wet swampy border under a wall facing to the south, and is enveloped at present in mud and mire!

Here is a large rookery round this house, the inhabitants of which seem to get their livelihood very easily; for they spend the greatest part of the day on their nest-trees when the weather is mild. These rooks retire every evening all the winter from this rookery, where they only call by the way, as they are going to roost in deep woods: at the dawn of day they always revisit their nest-trees, and are preceded a few minutes by a flight of daws, that act, as it were, as their harbingers.

I am, etc.

LETTER XVIII

Selborne, Jan. 29, 1774.

DEAR SIR,

The house-swallow, or chimney-swallow, is undoubtedly the first comer of all the British *hirundines*; and appears in general on or about the thirteenth of April, as I have remarked from many years' observation. Not but now and then a straggler is seen much earlier: and, in particular, when I was a boy I observed a swallow for a whole day together on a sunny warm Shrove Tuesday; which day could not fall out later than the middle of March, and often happened early in February.[130]

It is worth remarking that these birds are seen first about lakes and mill-ponds; and it is also very particular, that if these early visitors happen to find frost and snow, as was the case of the two dreadful springs of 1770 and 1771, they immediately withdraw for a time. A circumstance this much more in favour of hiding than migration; since it is much more probable that a bird should retire to its hybernaculum just at hand, than return for a week or two only to warmer latitudes.

The swallow, though called the chimney-swallow, by no means builds altogether in chimneys, but often within barns and out-houses against the rafters; and so she did in Virgil's time:

... Antè
Garrula quàm tignis nidos suspendat hirundo.

In Sweden she builds in barns, and is called *ladu swala*, the barn-swallow. Besides, in the warmer parts of Europe there are no chimneys to houses, except they are English-built: in these countries she constructs her nest in porches, and gateways, and galleries, and open halls.

Here and there a bird may affect some odd, peculiar place; as we have known a swallow build down the shaft of an old well, through which chalk had been formerly drawn up for the purpose of manure: but in general with us this *hirundo* breeds in chimneys;[131] and loves to haunt those stacks where there is a constant fire, no doubt for the sake of warmth. Not that it can subsist in the immediate shaft where there is a fire; but prefers one adjoining to that of the kitchen, and disregards the perpetual smoke of that funnel, as I have often observed with some degree of wonder.

Five or six or more feet down the chimney does this little bird begin to form her nest about the middle of May, which consists, like that of the house-martin, of a crust or shell composed of dirt or mud, mixed with short pieces of straw to render it tough and permanent; with this difference, that whereas the shell of the martin is nearly hemispheric, that of the swallow is open at the top, and like half a deep dish: this nest is lined with fine grasses, and feathers which are often collected as they float in the air.

Wonderful is the address which this adroit bird shows all day long in ascending and descending with security through so narrow a pass. When hovering over the mouth of the funnel, the vibrations of her wings acting on the confined air occasion a rumbling like thunder. It is not improbable that the dam submits to this inconvenient situation so low in the shaft, in order to secure her broods from rapacious birds, and particularly from owls, which frequently fall down chimneys, perhaps in attempting to get at these nestlings.

The swallow lays from four to six white eggs, dotted with red specks; and brings out her first brood about the last week in June, or the first week in July. The progressive method by which the young are introduced into life is very amusing: first, they emerge from the shaft with difficulty enough, and often fall down into the rooms below: for a day or so they are fed on the chimney-top, and then are conducted to the dead leafless bough of some tree, where, sitting in a row, they are attended with great assiduity, and may then be called perchers. In a day or two more they become flyers, but are still unable to take their own food; therefore they play about near the place where the dams are hawking for flies; and when a mouthful is collected, at a certain signal given, the dam and the nestling advance, rising towards each other, and meeting at an angle; the young one all the while uttering such a little quick note of gratitude and complacency, that a person must have paid very little regard to the wonders of nature that has not often remarked this feat.

The dam betakes herself immediately to the business of a second brood as soon as she is disengaged from her first; which at once associates with the first broods of house-martins; and with them congregates, clustering on sunny roofs, towers, and trees. This *hirundo* brings out her second brood towards the middle and end of August.

All the summer long is the swallow a most instructive pattern of unwearied industry and affection; for, from morning to night, while there is a family to be supported, she spends the whole day in skimming close to the ground, and exerting the most sudden turns and quick evolutions. Avenues, and long walks under hedges, and pasture-fields, and mown meadows where cattle graze, are her delight, especially if there are trees interspersed; because in such spots insects most abound. When a fly is taken a smart snap from her bill is heard, resembling the noise at the shutting of a watch-case; but the motion of the mandibles are too quick for the eye.

The swallow, probably the male bird, is the *excubitor* to house-martins, and other little birds, announcing the approach of birds of prey. For as soon as an hawk appears, with a shrill alarming note he calls all the swallows and martins about him; who pursue in a body, and buffet and strike their enemy till they have driven him from the village, darting down from above on his back, and rising in a perpendicular line in perfect security. This bird also will sound the alarm, and strike at cats when they climb on the roofs of houses, or otherwise approach the nests. Each species of *hirundo* drinks as it flies along, sipping the surface of the water; but the swallow alone, in general, washes on the wing, by dropping into a pool for many times together: in very hot weather house-martins and bank-martins dip and wash a little.

The swallow is a delicate songster, and in soft sunny weather sings both perching and flying; on trees in a kind of concert, and on chimney-tops: is also a bold flyer, ranging to distant downs and commons even in windy weather, which the other species seem much to dislike; nay, even frequenting exposed sea-port towns, and making little excursions over the salt water. Horsemen on wide downs are often closely attended by a little party of swallows for miles together, which plays before and behind them, sweeping around, and collecting all the skulking insects that are roused by the trampling of the horses' feet: when the wind blows hard, without this expedient, they are often forced to settle to pick up their lurking prey.

This species feeds much on little *coleoptera*, as well as on gnats and flies: and often settles on dug ground, or paths, for gravels to grind and digest its food. Before they depart, for some weeks, to a bird, they forsake houses and chimnies, and roost in trees; and usually withdraw about the beginning of October; though some few stragglers may appear on at times till the first week in November.

Some few pairs haunt the new and open streets of London next the fields, but do not enter, like the house-martin, the close and crowded parts of the city.

Both male and female are distinguished from their congeners by the length and forkedness of their tails. They are undoubtedly the most nimble of all the species: and when the male pursues the female in amorous chase, they then go beyond their usual speed, and exert a rapidity almost too quick for the eye to follow.

After this circumstantial detail of the life and discerning στοργὴ of the swallow, I shall add, for your farther amusement, an anecdote or two not much in favour of her sagacity:

A certain swallow built for two years together on the handles of a pair of garden-shears, that were stuck up against the boards in an out-house, and therefore must have her nest spoiled whenever that implement was wanted: and, what is stranger still, another bird of the same species built its nest on the wings and body of an owl that happened by accident to hang dead and dry from the rafter of a barn. This owl, with the nest on its wings, and with eggs in the nest, was brought as a curiosity worthy the most elegant private museum in Great Britain. The owner, struck with the oddity of the sight, furnished the bringer with a large shell, or conch, desiring him to fix it just where the owl hung: the person did as he was ordered, and the following year a pair, probably the same pair, built their nest in the conch, and laid their eggs.

The owl and the conch make a strange grotesque appearance, and are not the least curious specimens in that wonderful collection of art and nature.*

Thus is instinct in animals, taken the least out of its way, an undistinguishing, limited faculty; and blind to every circumstance that does not immediately respect self-preservation, or lead at once to the propagation or support of their species.

I am,
With all respect, etc., etc.

LETTER XIX

Selborne, Feb. 14, 1774.

DEAR SIR,

I received your favour of the eighth, and am pleased to find that you read my little history of the swallow with your usual candour: nor was I less pleased to find that you made objections where you saw reason.

As to the quotations, it is difficult to say precisely which species of *hirundo* Virgil might intend in the lines in question, since the ancients did not attend to specific differences like modern naturalists: yet somewhat may be gathered, enough to incline me to suppose that in the two passages quoted the poet had his eye on the swallow.

In the first place the epithet *garrula* suits the swallow well,

*Sir Ashton Lever's Museum.

who is a great songster; but not the martin, which is rather a mute bird; and when it sings is so inward as scarce to be heard. Besides, if *tignum* in that place signifies a rafter rather than a beam, as it seems to me to do, then I think it must be the swallow that is alluded to, and not the martin; since the former does frequently build within the roof against the rafters; while the latter always, as far as I have been able to observe, builds without the roof against eaves and cornices.

As to the simile, too much stress must not be laid on it: yet the epithet *nigra* speaks plainly in favour of the swallow, whose back and wings are very black; while the rump of the martin is milk-white, its back and wings blue, and all its under part white as snow. Nor can the clumsy motions (comparatively clumsy) of the martin well represent the sudden and artful evolutions and quick turns which Juturna gave to her brother's chariot, so as to elude the eager pursuit of the enraged Æneas. The verb *sonat* also seems to imply a bird that is somewhat loquacious.*

We have had a very wet autumn and winter, so as to raise the springs to a pitch beyond any thing since 1764; which was a remarkable year for floods and high waters. The land-springs, which we call *lavants*, break out much on the downs of Sussex, Hampshire, and Wiltshire. The country people say when the lavants rise corn will always be dear; meaning that when the earth is so glutted with water as to send forth springs on the downs and uplands, that the corn-vales must be drowned; and so it has proved for these ten or eleven years past. For land-springs have never obtained more since the memory of man than during that period; nor has there been known a greater scarcity of all sorts of grain, considering the great improvement

---

* *Nigra* velut magnas domini cum divitis ædes
Pervolat, et pennis alta atria lustrat hirundo,
Pabula parva legens, nidisque loquacibus escas:
Et nunc porticibus vacuis, nunc humida circum
Stagna *sonat*. . . .

of modern husbandry. Such a run of wet seasons a century or two ago would, I am persuaded, have occasioned a famine. Therefore pamphlets and newspaper letters, that talk of combinations, tend to inflame and mislead; since we must not expect plenty till Providence sends us more favourable seasons.

The wheat of last year, all round this district, and in the county of Rutland, and elsewhere, yields remarkably bad: and our wheat on the ground, by the continual late sudden vicissitudes from fierce frost to pouring rains, looks poorly; and the turnips rot very fast.

LETTER XX

Selborne, Feb. 26, 1774.

DEAR SIR,

The sand-martin, or bank-martin, is by much the least of any of the British *hirundines*; and, as far as we have ever seen, the smallest known *hirundo*; though Brisson asserts that there is one much smaller, and that is the *hirundo esculenta*.[132]

But it is much to be regretted that it is scarce possible for any observer to be so full and exact as he could wish in reciting the circumstances attending the life and conversation of this little bird, since it is *fera naturâ*, at least in this part of the kingdom, disclaiming all domestic attachments, and haunting wild heaths and commons where there are large lakes; while the other species, especially the swallow and house-martin, are remarkably gentle and domesticated, and never seem to think themselves safe but under the protection of man.

Here are in this parish, in the sand-pits and banks of the lakes of Wolmer-forest, several colonies of these birds; and yet they are never seen in the village; nor do they at all frequent the cottages that are scattered about in that wild district. The only instance I ever remember where this species haunts any building is at the town of Bishop's Waltham, in this county, where many sand-martins nestle and breed in the scaffold-holes of the back-wall of William of Wykeham's stables: but then this wall stands in a very sequestered and retired enclosure, and faces upon a large and beautiful lake. And indeed this species seems so to delight in large waters, that no instance occurs of their abounding, but near vast pools or rivers: and in particular it has been remarked that they swarm in the banks of the Thames in some places below London-bridge.

It is curious to observe with what different degrees of architectonic skill Providence has endowed birds of the same genus, and so nearly correspondent in their general mode of life! for while the swallow and the house-martin discover the greatest address in raising and securely fixing crusts or shells of loam as *cunabula* for their young, the bank-martin terebrates a round and regular hole in the sand or earth, which is serpentine, horizontal, and about two feet deep. At the inner end of this burrow does this bird deposit, in a good degree of safety, her rude nest, consisting of fine grasses and feathers, usually goose-feathers, very inartificially laid together.

Perseverance will accomplish anything: though at first one would be disinclined to believe that this weak bird, with her soft and tender bill and claws, should ever be able to bore the stubborn sand-bank without entirely disabling herself; yet with these feeble instruments have I seen a pair of them make great dispatch: and could remark how much they had scooped that day by the fresh sand which ran down the bank, and was of a different colour from that which lay loose and bleached in the sun.

In what space of time these little artists are able to mine and finish these cavities I have never been able to discover, for reasons given above; but it would be a matter worthy of obser-vation, where it falls in the way of any naturalist to make his remarks. This I have often taken notice of, that several holes of different depths are left unfinished at the end of summer. To imagine that these beginnings were intentionally made in order to be in the greater forwardness for next spring, is allowing perhaps too much foresight and *rerum prudentia* to a simple bird. May not the cause of these *latebræ* being left unfinished arise from their meeting in those places with strata too harsh, hard, and solid, for their purpose, which they relinquish, and go to a fresh spot that works more freely? Or may they not in other places fall in with a soil as much too loose and mouldering, liable to flounder, and threatening to overwhelm them and their labours?

One thing is remarkable – that, after some years, the old holes are forsaken and new ones bored; perhaps because the old habitations grow foul and fetid from long use, or because they may so abound with fleas as to become untenable. This species of swallow moreover is strangely annoyed with fleas: and we have seen fleas, bed-fleas (*pulex irritans*),[133] swarming at the mouths of these holes, like bees upon the stools of their hives.

The following circumstance should by no means be omitted – that these birds do *not* make use of their caverns by way of *hybernacula*, as might be expected; since banks so perforated have been dug out with care in the winter, when nothing was found but empty nests.

The sand-martin arrives much about the same time with the swallow, and lays, as she does, from four to six white eggs. But as the species is *cryptogame*, carrying on the business of nidification, incubation, and the support of its young in the dark, it would not be so easy to ascertain the time of breeding, were it not for the coming forth of the broods, which appear much about the time, or rather somewhat earlier than those of the swallow. The nestlings are supported in common like those of their congeners, with gnats and other small insects; and sometimes they are fed with *libellulæ* (dragon-flies) almost as long as themselves. In the last week in June we have seen a row of these sitting on a rail near a great pool as perchers; and so young and helpless, as easily to be taken by hand: but whether the dams ever feed them on the wing, as swallows and house-martins do, we have never yet been able to determine; nor do we know whether they pursue and attack birds of prey.

When they happen to breed near hedges and enclosures, they are dispossessed of their breeding holes by the house-sparrow, which is on the same account a fell adversary to house-martins.

These *hirundines* are no songsters, but rather mute,[134] making only a little harsh noise when a person approaches their nests. They seem not to be of a sociable turn, never with us congregating with their congeners in the autumn. Undoubtedly they breed a second time, like the house-martin and swallow; and withdraw about Michaelmas.

Though in some particular districts they may happen to abound, yet in the whole, in the south of England at least, is this much the rarest species. For there are few towns or large villages but what abound with house-martins; few churches, towers, or steeples, but what are haunted by some swifts; scarce a hamlet or single cottage-chimney that has not its swallow; while the bank-martins, scattered here and there, live a sequestered life among some abrupt sand-hills, and in the banks of some few rivers.

These birds have a peculiar manner of flying; flitting about with odd jerks, and vacillations, not unlike the motions of a butterfly. Doubtless the flight of all *hirundines* is influenced by, and adapted to, the peculiar sort of insects which furnish their food. Hence it would be worth inquiry to examine what particular group of insects affords the principal food of each respective species of swallow.

Notwithstanding what has been advanced above, some few sand-martins, I see, haunt the skirts of London, frequenting the dirty pools in Saint George's-Fields, and about White-Chapel. The question is where these build, since there are no banks or bold shores in that neighbourhood: perhaps they nestle in the scaffold-holes of some old or new deserted building. They dip and wash as they fly sometimes, like the house-martin and swallow.

Sand-martins differ from their congeners in the diminutiveness of their size, and in their colour, which is what is usually called a mouse-colour. Near Valencia in Spain, they are taken, says Willughby, and sold in the markets for the table; and are called by the country people, probably from their desultory jerking manner of flight, *Papilion de montagna*.

LETTER XXI

Selborne, Sept. 28, 1774.

Dear Sir,

As the swift or black-martin is the largest of the British *hirundines*, so is it undoubtedly the latest comer. For I remember but one instance of its appearing before the last week in April: and in some of our late frosty, harsh springs, it has not been seen till the beginning of May. This species usually arrives in pairs.

The swift, like the sand-martin, is very defective in architecture, making no crust, or shell, for its nest; but forming it of dry grasses and feathers, very rudely and inartificially put together. With all my attention to these birds, I have never been able once to discover one in the act of collecting or carrying in materials: so that I have suspected (since their nests are exactly the same) that they sometimes usurp upon the house-sparrows, and expel them, as sparrows do the house and sand-martin; well remembering that I have seen them squabbling together at the entrance of their holes; and the sparrows up in arms, and much disconcerted at these intruders. And yet I am assured, by a nice observer[135] in such matters, that they do collect feathers for their nests in Andalusia; and that he has shot them with such materials in their mouths.

Swifts, like sand-martins, carry on the business of nidification quite in the dark, in crannies of castles, and towers, and steeples, and upon the tops of the walls of churches under the roof; and therefore cannot be so narrowly watched as those species that build more openly: but, from what I could ever observe, they begin nesting about the middle of May; and I have remarked, from eggs taken, that they have sat hard by the ninth of June. In general they haunt tall buildings, churches, and steeples, and breed only in such: yet in this village some pairs frequent the lowest and meanest cottages, and educate their young under those thatched roofs. We remember but one instance where they breed out of buildings; and that is in the sides of a deep chalk-pit near the town of Odiham, in this county, where we have seen many pairs entering the crevices, and skimming and squeaking round the precipices.

As I have regarded these amusive birds with no small attention, if I should advance something new and peculiar with respect to them, and different from all other birds, I might perhaps be credited; especially as my assertion is the result of many years' exact observation. The fact that I would advance is, that swifts tread, or copulate, on the wing: and I would wish any nice observer, that is startled at this supposition, to use his own eyes, and I think he will soon be convinced. In another class of animals, viz., the insect, nothing is so common as to see the different species of many genera in conjunction as they fly. The swift is almost continually on the wing; and as it never settles on the ground, on trees, or roofs, would seldom find opportunity for amorous rites, was it not enabled to indulge them in the air. If any person would watch these birds of a fine morning in May, as they are sailing round at a great height from the ground, he would see, every now and then, one drop on the back of another, and both of them sink down together for many fathoms with a loud piercing shriek. This I take to be the juncture when the business of generation is carrying on.

As the swift eats, drinks, collects materials for its nest, and, as it seems, propagates on the wing; it appears to live more in the air than any other bird, and to perform all functions there save those of sleeping and incubation.

This *hirundo* differs widely from its congeners in laying invariably but two eggs at a time,[136] which are milk-white, long, and peaked at the small end; whereas the other species lay at each brood from four to six. It is a most alert bird, rising very early, and retiring to roost very late; and is on the wing in the height of summer at least sixteen hours. In the longest days it does not withdraw to rest till a quarter before nine in the evening, being the latest of all day birds. Just before they retire whole groups of them assemble high in the air, and squeak, and shoot about with wonderful rapidity. But this bird is never so much alive as in sultry thundry weather, when it expresses great alacrity, and calls forth all its powers. In hot mornings several, getting together in little parties, dash round the steeples and churches, squeaking as they go in a very clamorous manner; these, by nice observers, are supposed to be males, serenading their sitting hens; and not without reason, since they seldom squeak till they come close to the walls or eaves, and since those within utter at the same time a little inward note of complacency.

When the hen has sat hard all day, she rushes forth just as it is almost dark, and stretches and relieves her weary limbs, and snatches a scanty meal for a few minutes, and then returns to her duty of incubation. Swifts, when wantonly and cruelly shot while they have young, discover a little lump of insects in their mouths, which they pouch and hold under their tongue. In general they feed in a much higher district than the other species; a proof that gnats and other insects do also abound to a considerable height in the air: they also range to vast distances; since locomotion is no labour to them, who are endowed with such wonderful powers of wing. Their powers seem to be in proportion to their levers; and their wings are longer in proportion than those of almost any other bird. When they mute, or ease themselves in flight, they raise their wings, and make them meet over their backs.

At some certain times in the summer I had remarked that swifts were hawking very low for hours together over pools and streams; and could not help inquiring into the object of their pursuit that induced them to descend so much below their usual range. After some trouble, I found that they were taking *phryganeæ, ephemeræ,* and *libellulæ* (cadew-flies, may-flies, and dragon-flies) that were just emerged out of their aurelia state. I then no longer wondered that they should be so willing to stoop for a prey that afforded them such plentiful and succulent nourishment.

They bring out their young about the middle or latter end of July: but as these never become perchers, nor, that ever I could discern, are fed on the wing by their dams, the coming forth of the young is not so notorious as in the other species.

On the thirtieth of last June I untiled the eaves of an house where many pairs build, and found in each nest only two squab naked *pulli*: on the eighth of July I repeated the same inquiry, and found they had made very little progress towards a fledged state, but were still naked and helpless. From whence we may conclude that birds whose way of life keeps them perpetually on the wing would not be able to quit their nest till the end of the month. Swallows and martins, that have numerous families, are continually feeding them every two or three minutes; while swifts, that have but two young to maintain, are much at their leisure, and do not attend on their nests for hours together.

Sometimes they pursue and strike at hawks that come in their way; but not with that vehemence and fury that swallows express on the same occasion. They are out all day long in wet days, feeding about, and disregarding still rain: from whence two things may be gathered; first, that many insects abide high in the air, even in rain; and next, that the feathers of these birds must be well preened to resist so much wet. Windy, and particularly windy weather with heavy showers, they dislike; and on such days withdraw, and are scarce ever seen.

There is a circumstance respecting the colour of swifts, which seems not to be unworthy our attention. When they arrive in the spring they are all over of a glossy, dark soot-colour, except their chins, which are white; but, by being all day long in the sun and air, they become quite weather-beaten and bleached before they depart, and yet they return glossy again in the spring. Now, if they pursue the sun into lower latitudes, as some suppose, in order to enjoy a perpetual summer, why do they not return bleached? Do they not rather perhaps retire to rest for a season,[137] and at that juncture moult and change their feathers, since all other birds are known to moult soon after the season of breeding?

Swifts are very anomalous in many particulars, dissenting from all their congeners not only in the number of their young, but in breeding but once in a summer; whereas all the other British *hirundines* breed invariably twice. It is past all doubt that swifts can breed but once, since they withdraw in a short time after the flight of their young, and some time before their congeners bring out their second brood. We may here remark, that, as swifts breed but once in a summer, and only two at a time, and the other *hirundines* twice, the latter, who lay from four to six eggs, increase at an average five times as fast as the former.

But in nothing are swifts more singular than in their early retreat. They retire, as to the main body of them, by the tenth of August, and sometimes a few days sooner: and every straggler invariably withdraws by the twentieth, while their congeners, all of them, stay till the beginning of October; many of them all through that month, and some occasionally to the beginning of November. This early retreat is mysterious and wonderful, since that time is often the sweetest season in the year. But, what is more extraordinary, they begin to retire still earlier in the most southerly parts of Andalusia, where they can be no ways influenced by any defect of heat; or, as one might suppose, defect of food. Are they regulated in their motions with us by a failure of food, or by a propensity to moulting, or by a disposition to rest after so rapid a life, or by what? This is one of those incidents in natural history that not only baffles our searches, but almost eludes our guesses!

These *hirundines* never perch on trees or roofs, and so never congregate with their congeners. They are fearless while haunting their nesting places, and are not to be scared with a gun; and are often beaten down with poles and cudgels as they stoop to go under the eaves. Swifts are much infested with those pests to the genus called *hippoboscæ hirundinis*; and often wriggle and scratch themselves, in their flight, to get rid of that clinging annoyance.

Swifts are no songsters, and have only one harsh screaming note; yet there are ears to which it is not displeasing, from an aggreeable association of ideas, since that note never occurs but in the most lovely summer weather.

They never settle on the ground but through accident; and when down can hardly rise, on account of the shortness of their legs and the length of their wings: neither can they walk, but only crawl; but they have a strong grasp with their feet, by which they cling to walls. Their bodies being flat they can enter a very narrow crevice; and where they cannot pass on their bellies they will turn up edgewise.

The particular formation of the foot discriminates the swift from all British *hirundines*; and indeed from all other known birds, the *hirundo melba*, or great white-bellied swift of Gibraltar, excepted; for it is so disposed as to carry '*omnes quatuor digitos anticos*' all its four toes forward; besides, the least toe, which should be the back-toe, consists of one bone alone, and the other three only of two apiece. A construction most rare and peculiar, but nicely adapted to the purposes in which their feet are employed. This, and some peculiarities attending the nostrils and under mandible, have induced a discerning naturalist* to suppose that this species might constitute a *genus per se*.

In London a party of swifts frequents the Tower, playing and feeding over the river just below the bridge; others haunt some of the churches of the Borough next the fields; but do not venture, like the house-martin, into the close crowded part of the town.

The Swedes have bestowed a very pertinent name on this swallow, calling it *ring swala*, from the perpetual rings or circles that it takes round the scene of its nidification.

Swifts feed on *coleoptera*, or small beetles with hard cases over their wings, as well as on the softer insects; but it does not appear how they can procure gravel to grind their food, as swallows do, since they never settle on the ground. Young ones, over-run with *hippoboscæ*, are sometimes found, under their nests, fallen to the ground: the number of vermin rendering their abode insupportable any longer. They frequent in this village several abject cottages: yet a succession still haunts the same unlikely roofs: a good proof this that the same birds return to the same spots. As they must stoop very low to get up under these humble eaves, cats lie in wait, and sometimes catch them on the wing.

On the fifth of July, 1775, I again untiled part of a roof over the nest of a swift. The dam sat in the nest; but so strongly was she affected by natural στοργὴ for her brood, which she supposed to be in danger, that, regardless of her own safety, she would not stir, but lay sullenly by them, permitting herself to be taken in hand. The squab young we brought down and placed on the grass-plot, where they tumbled about, and were as helpless as a new-born child. While we contemplated their naked bodies, their unwieldy disproportioned *abdomina*, and their heads, too heavy for their necks to support, we could not but wonder when we reflected that these shiftless beings in a little more than a fortnight would be able to dash through the air almost with the inconceivable swiftness of a meteor; and per-

*John Antony Scopoli, of Carniola, M.D.

haps, in their emigration must traverse vast continents and oceans as distant as the equator. So soon does nature advance small birds to their ἡλικία, or state of perfection; while the progressive growth of men and large quadrupeds is slow and tedious!

I am, etc.

LETTER XXII

Selborne, Sept. 13, 1774.

DEAR SIR,

By means of a straight cottage chimney I had an opportunity this summer of remarking, at my leisure, how swallows ascend and descend through the shaft; but my pleasure, in contemplating the address with which this feat was performed to a considerable depth in the chimney, was somewhat interrupted by apprehensions lest my eyes might undergo the same fate with those of Tobit.*[138]

Perhaps it may be some amusement to you to hear at what times the different species of *hirundines* arrived this spring in three very distant counties of this kingdom. With us the swallow was seen first on April the 4th, the swift on April the 24th, the bank-martin on April the 12th, and the house-martin not till April the 30th. At South Zele, Devonshire, swallows did not arrive till April the 25th; swifts, in plenty, on May the 1st; and house-martins not till the middle of May. At Blackburn, in Lancashire, swifts were seen April the 28th, swallows April the 29th, house-martins May the 1st. Do these different dates, in such distant districts, prove anything for or against migration?

A farmer, near Weyhill, fallows his land with two teams of asses; one of which works till noon, and the other in the after-

*Tobit ii. 10.

noon. When these animals have done their work, they are penned, all night, like sheep, on the fallow. In the winter they are confined and foddered in a yard, and make plenty of dung.

Linnæus says that hawks *paciscuntur inducias cum avibus, quamdiu cuculus cuculat*:[139] but it appears to me that, during that period, many little birds are taken and destroyed by birds of prey, as may be seen by their feathers left in lanes and under hedges.

The missel-thrush is, while breeding, fierce and pugnacious, driving such birds as approach its nest, with great fury, to a distance. The Welch call it *pen y llwyn*, the head or master of the coppice. He suffers no magpie, jay, or blackbird, to enter the garden where he haunts; and is, for the time, a good guard to the new-sown legumens. In general he is very successful in the defence of his family: but once I observed in my garden, that several magpies came determined to storm the nest of a missel-thrush: the dams defended their mansion with great vigour, and fought resolutely *pro aris & focis*; but numbers at last prevailed, they tore the nest to pieces, and swallowed the young alive.

In the season of nidification the wildest birds are comparatively tame. Thus the ring-dove breeds in my fields, though they are continually frequented; and the missel-thrush, though most shy and wild in the autumn and winter, builds in my garden close to a walk where people are passing all day long.

Wall-fruit abounds with me this year: but my grapes, that used to be forward and good, are at present backward beyond all precedent: and this is not the worst of the story; for the same ungenial weather, the same black cold solstice, has injured the more necessary fruits of the earth, and discoloured and blighted our wheat. The crop of hops promises to be very large.

Frequent returns of deafness incommode me sadly, and half disqualify me for a naturalist; for, when those fits are upon me, I lose all the pleasing notices and little intimations arising from rural sounds: and May is to me as silent and mute with respect to the notes of birds, etc., as August. My eyesight is, thank God, quick and good; but with respect to the other sense, I am, at times, disabled:

> And Wisdom at one entrance quite shut out.

LETTER XXIII

Selborne, June 8, 1775.

DEAR SIR,

On September the 21st, 1741, being then on a visit, and intent on field-diversions, I rose before daybreak: when I came into the enclosures, I found the stubbles and clover-grounds matted all over with a thick coat of cobweb, in the meshes of which a copious and heavy dew hung so plentifully that the whole face of the country seemed, as it were, covered with two or three setting-nets drawn one over another. When the dogs attempted to hunt, their eyes were so blinded and hoodwinked that they could not proceed, but were obliged to lie down and scrape the incumbrances from their faces with their fore-feet, so that, finding my sport interrupted, I returned home musing in my mind on the oddness of the occurrence.

As the morning advanced the sun became bright and warm, and the day turned out one of those most lovely ones which no season but the autumn produces: cloudless, calm, serene, and worthy of the South of France itself.

About nine an appearance very unusual began to demand our attention, a shower of cobwebs falling from very elevated regions, and continuing, without any interruption, till the close of the day. These webs were not single filmy threads, floating in the air in all directions, but perfect flakes or rags; some near an inch broad, and five or six long, which fell with a degree of velocity which showed they were considerably heavier than the atmosphere.

On every side as the observer turned his eyes might he behold a continual succession of fresh flakes falling into his sight, and twinkling like stars as they turned their sides towards the sun.

How far this wonderful shower extended would be difficult to say; but we know that it reached Bradley, Selborne, and Alresford, three places which lie in a sort of a triangle, the shortest of whose sides is about eight miles in extent.

At the second of those places there was a gentleman[140] (for whose veracity and intelligent turn we have the greatest veneration) who observed it the moment he got abroad; but concluded that, as soon as he came upon the hill above his house, where he took his morning rides, he should be higher than this meteor, which he imagined might have been blown, like thistle-down, from the common above: but, to his great astonishment, when he rode to the most elevated part of the down, 300 feet above his fields, he found the webs in appearance still as much above him as before; still descending into sight in a constant succession, and twinkling in the sun, so as to draw the attention of the most incurious.

Neither before nor after was any such fall observed; but on this day the flakes hung in the trees and hedges so thick, that a diligent person sent out might have gathered baskets full.

The remark that I shall make on these cobweb-like appearances, called gossamer, is, that, strange and superstitious as the notions about them were formerly, nobody in these days doubts but that they are the real production of small spiders, which swarm in the fields in fine weather in autumn, and have a power of shooting out webs from their tails so as to render themselves buoyant, and lighter than air. But why these apterous insects should that day take such a wonderful aerial excursion, and why their webs should at once become so gross and material as to be considerably more weighty than air, and to descend with precipitation, is a matter beyond my skill. If I might be allowed to hazard a supposition, I should imagine that those filmy threads, when first shot, might be entangled in the rising dew, and so drawn up, spiders and all, by a brisk evaporation into the region where clouds are formed: and if the spiders have a power of coiling and thickening their webs in the air, as Dr Lister[141] says they have [see his Letters to Mr Ray], then, when they were become heavier than the air, they must fall.

Every day in fine weather, in autumn chiefly, do I see those spiders shooting out their webs and mounting aloft: they will go off from your finger if you will take them into your hand. Last summer one alighted on my book as I was reading in the parlour; and, running to the top of the page, and shooting out a web, took its departure from thence. But what I most wondered at, was that it went off with considerable velocity in a place where no air was stirring; and I am sure that I did not assist it with my breath. So that these little crawlers seem to have while mounting, some loco-motive power without the use of wings, and to move in the air, faster than the air itself.

LETTER XXIV

Selborne, Aug. 15, 1775.

DEAR SIR,

There is a wonderful spirit of sociality in the brute creation, independent of sexual attachment: the congregating of gregarious birds in the winter is a remarkable instance.

Many horses, though quiet with company, will not stay one minute in a field by themselves: the strongest fences cannot restrain them. My neighbour's horse will not only not stay by himself abroad, but he will not bear to be left alone in a strange stable without discovering the utmost impatience, and endeavouring to break the rack and manger with his fore feet. He has been known to leap out at a stable-window, through which dung was thrown, after company; and yet in other respects is remarkably quiet. Oxen and cows will not fatten by themselves but will neglect the finest pasture that is not recommended by society. If would be needless to instance in sheep, which constantly flock together.

But this propensity seems not to be confined to animals of the same species; for we know a doe still alive, that was brought up from a little fawn with a dairy of cows; with them it goes afield, and with them it returns to the yard. The dogs of the house take no notice of this deer, being used to her; but, if strange dogs come by, a chase ensues; while the master smiles to see his favourite securely leading her pursuers over hedge, or gate, or stile, till she returns to the cows, who, with fierce lowings and menacing horns, drive the assailants quite out of the pasture.

Even great disparity of kind and size does not always prevent social advances and mutual fellowship. For a very intelligent and observant person has assured me that, in the former part of his life, keeping but one horse, he happened also on a time to have but one solitary hen. These two incongruous animals spent much of their time together in a lonely orchard, where they saw no creature but each other. By degrees an apparent regard began to take place between these two sequestered individuals. The fowl would approach the quadruped with notes of complacency, rubbing herself gently against his legs; while the horse would look down with satisfaction, and move with the greatest caution and circumspection, lest he should trample on his diminutive companion. Thus, by mutual good offices, each seemed to console the vacant hours of the other: so that Milton, when he puts the following sentiment in the mouth of Adam, seems to be somewhat mistaken:

> Much less can bird with beast, or fish with fowl,
> So well converse, nor with the ox the ape.

I am, etc.

LETTER XXV

Selborne, Oct. 2, 1775.

DEAR SIR,

We have two gangs or hordes of gypsies which infest the south and west of England, and come round in their circuit two or three times in the year. One of these tribes calls itself by the noble name of Stanley, of which I have nothing particular to say; but the other is distinguished by an appellative somewhat remarkable. – As far as their harsh gibberish can be understood, they seem to say that the name of their clan is Curleople; now the termination of this word is apparently Grecian: and as Mezeray[142] and the gravest historians all agree that these vagrants did certainly migrate from Egypt and the East two or three centuries ago, and so spread by degrees over Europe, may not this name, a little corrupted, be the very name they brought with them from the Levant? It would be matter of some curiosity, could one meet with an intelligent person among them, to inquire whether, in their jargon, they still retain any Greek words: the Greek radicals will appear in hand, foot, head, water, earth, etc. It is possible that amidst their cant and corrupted dialect many mutilated remains of their native language might still be discovered.

With regard to those peculiar people, the gypsies, one thing is very remarkable, and especially as they came from warmer climates; and that is, that while other beggars lodge in barns, stables, and cow-houses, these sturdy savages seem to pride themselves in braving the severities of winter, and in living *sub dio* the whole year round. Last September was as wet a month as ever was known; and yet during those deluges did a young gypsy-girl lie-in in the midst of one of our hop-gardens, on the cold ground, with nothing over her but a piece of blanket extended on a few hazel-rods bent hoop-fashion, and stuck into the earth at each end, in circumstances too trying for a cow in the same condition: yet within this garden there was a large hop-kiln, into the chambers of which she might have retired, had she thought shelter an object worthy her attention.

Europe itself, it seems, cannot set bounds to the rovings of those vagabonds; for Mr Bell, in his return from Peking, met a gang of these people on the confines of Tartary, who were endeavouring to penetrate those deserts and try their fortune in China.*

Gypsies are called in French, Bohemians; in Italian and modern Greek, Zingari.

I am, etc.

LETTER XXVI

Selborne, Nov. 1, 1775.

DEAR SIR,

> Hic ... tædæ pingues, hic plurimus ignis
> Semper, et assiduâ postes fuligine nigri.

I shall make no apology for troubling you with the detail of a very simple piece of domestic œconomy, being satisfied that you think nothing beneath your attention that tends to utility: the matter alluded to is the use of rushes instead of candles, which I am well aware prevails in many districts besides this; but as I know there are countries also where it does not obtain, and as I have considered the subject with some degree of exactness, I shall proceed in my humble story, and leave you to judge of the expediency.

The proper species of rush for this purpose seems to be the *juncus effusus*, or common soft rush, which is to be found in most

*See Bell's *Travels in China.*

moist pastures, by the sides of streams, and under hedges. These rushes are in best condition in the height of summer; but may be gathered, so as to serve the purpose well, quite on to autumn. It would be needless to add that the largest and longest are best. Decayed labourers, women, and children, make it their business to procure and prepare them. As soon as they are cut they must be flung into water, and kept there; for otherwise they will dry and shrink, and the peel will not run. At first a person would find it no easy matter to divest a rush of its peel or rind, so as to leave one regular, narrow, even rib from top to bottom that may support the pith: but this, like other feats, soon becomes familiar even to children; and we have seen an old woman, stone-blind, performing this business with great dispatch, and seldom failing to strip them with the nicest regularity. When these *junci* are thus far prepared, they must lie out on the grass to be bleached, and take the dew for some nights, and afterwards be dried in the sun.

Some address is required in dipping these rushes in the scalding fat or grease; but this knack also is to be attained by practice. The careful wife of an industrious Hampshire labourer obtains all her fat for nothing; for she saves the scummings of her bacon-pot for this use; and, if the grease abounds with salt, she causes the salt to precipitate to the bottom, by setting the scummings in a warm oven. Where hogs are not much in use, and especially by the sea-side, the coarser animal oils will come very cheap. A pound of common grease may be procured for four pence; and about six pounds of grease will dip a pound of rushes; and one pound of rushes may be bought for one shilling: so that a pound of rushes, medicated and ready for use, will cost three shillings. If men that keep bees will mix a little wax with the grease, it will give it a consistency, and render it more cleanly, and make the rushes burn longer: mutton-suet would have the same effect.

A good rush, which measured in length two feet four inches and an half, being minuted, burnt only three minutes short of an hour: and a rush still of greater length has been known to burn one hour and a quarter.

These rushes give a good clear light. Watch-lights (coated with tallow), it is true, shed a dismal one, 'darkness visible'; but then the wicks of those have two ribs of the rind, or peel, to support the pith, while the wick of the dipped rush has but one. The two ribs are intended to impede the progress of the flame, and make the candle last.

In a pound of dry rushes, avoirdupois, which I caused to be weighed and numbered, we found upwards of one thousand six hundred individuals. Now suppose each of these burns, one with another, only half an hour, then a poor man will purchase eight hundred hours of light, a time exceeding thirty-three entire days, for three shillings. According to this account each rush, before dipping, costs $\frac{1}{33}$ of a farthing, and $\frac{1}{11}$ afterwards. Thus a poor family will enjoy $5\frac{1}{2}$ hours of comfortable light for a farthing. An experienced old housekeeper assures me that one pound and a half of rushes completely supplies his family the year round, since working people burn no candle in the long days, because they rise and go to bed by daylight.

Little farmers use rushes much in the short days, both morning and evening in the dairy and kitchen; but the very poor, who are always the worst œconomists, and therefore must continue very poor, buy an halfpenny candle every evening, which, in their blowing open rooms, does not burn much more than two hours. Thus have they only two hours' light for their money instead of eleven.

While on the subject of rural œconomy, it may not be improper to mention a pretty implement of housewifery that we have seen no where else; that is, little neat besoms which our foresters make from the stalk of the *polytricum commune*, or great golden maiden-hair, which they call silk-wood, and find plenty in the bogs. When this moss is well combed and dressed, and divested of its outer skin, it becomes of a beautiful bright chestnut colour; and, being soft and pliant, is very proper for the dusting of beds, curtains, carpets, hangings, etc. If these besoms were known to the brushmakers in town, it is probable they might come much in use for the purpose above-mentioned.*

I am, etc.

LETTER XXVII

Selborne, December 12, 1775.

DEAR SIR,

We had in this village more than twenty years ago an idiot-boy, whom I well remember, who, from a child, showed a strong propensity to bees; they were his food, his amusement, his sole object. And as people of this cast have seldom more than one point in view, so this lad exerted all his few faculties on this one pursuit. In the winter he dosed away his time, within his father's house, by the fireside, in a kind of torpid state, seldom departing from the chimney-corner; but in the summer he was all alert, and in quest of his game in the fields, and on sunny banks. Honey-bees, humble-bees, and wasps, were his prey wherever he found them: he had no apprehensions from their stings, but would seize them *nudis manibus*, and at once disarm them of their weapons, and suck their bodies for the sake of their honey-bags. Sometimes he would fill his bosom between his shirt and his skin with a number of these captives; and sometimes would confine them in bottles. He was a very *merops apiaster*, or bee-bird; and very injurious to men that kept bees; for he would slide into their bee-gardens, and, sitting down before the stools, would rap with his finger on the hives, and so take the bees as they came out. He has been known to overturn hives for the sake of honey, of which he was passionately fond. Where metheglin[143] was making he would

*A besom of this sort is to be seen in Sir Ashton Lever's Museum.

linger round the tubs and vessels, begging a draught of what he
called bee-wine. As he ran about he used to make a humming
noise with his lips, resembling the buzzing of bees. This lad was
lean and sallow, and of a cadaverous complexion; and, except
in his favourite pursuit, in which he was wonderfully adroit,
discovered no manner of understanding. Had his capacity been
better, and directed to the same object, he had perhaps abated
much of our wonder at the feats of a more modern exhibitor
of bees; and we may justly say of him now,

> ... Thou,
> Had thy presiding star propitious shone,
> Should'st Wildman be...

When a tall youth he was removed from hence to a distant
village, where he died, as I understand, before he arrived at
manhood.

I am, etc.

LETTER XXVIII

Selborne, Jan. 8, 1776.

DEAR SIR,

It is the hardest thing in the world to shake off superstitious
prejudices: they are sucked in as it were with our mother's milk;
and growing up with us at a time when they take the fastest
hold and make the most lasting impressions, become so interwoven into our very constitutions, that the strongest good
sense is required to disengage ourselves from them. No wonder
therefore that the lower people retain them their whole lives
through, since their minds are not invigorated by a liberal
education, and therefore not enabled to make any efforts
adequate to the occasion.

Such a preamble seems to be necessary before we enter on the superstitions of this district, lest we should be suspected of exaggeration in a recital of practices too gross for this enlightened age.

But the people of Tring, in Hertfordshire, would do well to remember, that no longer ago than the year 1751, and within twenty miles of the capital, they seized on two superannuated wretches, crazed with age, and overwhelmed with infirmities, on a suspicion of witchcraft; and, by trying experiments, drowned them in a horse-pond.

In a farm-yard near the middle of this village stands, at this day, a row of pollard-ashes, which, by the seams and long cicatrices down their sides, manifestly show that, in former times, they have been cleft asunder. These trees, when young and flexible, were severed and held open by wedges, while ruptured children, stripped naked, were pushed through the apertures, under a persuasion that, by such a process, the poor babes would be cured of their infirmity. As soon as the operation was over, the tree, in the suffering part, was plastered with loam, and carefully swathed up. If the parts coalesced and soldered together, as usually fell out, where the feat was performed with any adroitness at all, the party was cured; but, where the cleft continued to gape, the operation, it was supposed, would prove ineffectual. Having occasion to enlarge my garden not long since, I cut down two or three such trees, one of which did not grow together.

We have several persons now living in the village, who, in their childhood, were supposed to be healed by this superstitious ceremony, derived down perhaps from our Saxon ancestors, who practised it before their conversion to Christianity.

At the south corner of the Plestor, or area, near the church, there stood, about twenty years ago, a very old grotesque hollow pollard-ash, which for ages had been looked on with no small veneration as a shrew-ash. Now a shrew-ash is an ash whose twigs or branches, when gently applied to the limbs of cattle, will immediately relieve the pains which a beast suffers from the running of a shrew-mouse over the part affected: for it is supposed that a shrew-mouse is of so baneful and deleterious a nature, that wherever it creeps over a beast, be it horse, cow, or sheep, the suffering animal is afflicted with cruel anguish, and threatened with the loss of the use of the limb. Against this accident, to which they were continually liable, our provident fore-fathers always kept a shrew-ash at hand, which, when once medicated, would maintain its virtue for ever. A shrew-ash was made thus:* Into the body of the tree a deep hole was bored with an auger, and a poor devoted shrew-mouse was thrust in alive, and plugged in, no doubt, with several quaint incantations long since forgotten. As the ceremonies necessary for such a consecration are no longer understood, all succession is at an end, and no such tree is known to subsist in the manor, or hundred.

As to that on the Plestor,

> The late vicar stubb'd and burnt it,

when he was way-warden, regardless of the remonstrances of the by-standers, who interceded in vain for its preservation, urging its power and efficacy, and alleging that it had been

> Religione patrum multos servata per annos.

I am, etc.

---

* For a similar practice, see Plot's *Staffordshire*.

LETTER XXIX

Selborne, Feb. 7, 1776.

DEAR SIR,

In heavy fogs, on elevated situations especially, trees are perfect alembics:[144] and no one that has not attended to such matters can imagine how much water one tree will distil in a night's time by condensing the vapour, which trickles down the twigs and boughs, so as to make the ground below quite in a float. In Newton-lane, in October 1775, on a misty day, a particular oak in leaf dropped so fast that the cart-way stood in puddles and the ruts ran with water, though the ground in general was dusty.

In some of our smaller islands in the West-Indies, if I mistake not, there are no springs or rivers; but the people are supplied with that necessary element, water, merely by the dripping of some large tall trees, which, standing in the bosom of a mountain, keep their heads constantly enveloped with fogs and clouds, from which they dispense their kindly never-ceasing moisture; and so render those districts habitable by condensation alone.

Trees in leaf have such a vast proportion more of surface than those that are naked, that, in theory, their condensations should greatly exceed those that are stripped of their leaves; but, as the former imbibe also a great quantity of moisture, it is difficult to say which drip most: but this I know, that deciduous trees that are entwined with much ivy seem to distil the greatest quantity. Ivy-leaves are smooth, and thick, and cold, and therefore condense very fast; and besides evergreens imbibe very little. These facts may furnish the intelligent with hints concerning what trees they should plant round small ponds that they would wish to be perennial; and show them how advantageous some trees are in preference to others.

Trees perspire profusely, condense largely, and check evapo-

ration so much, that woods are always moist: no wonder therefore that they contribute much to pools and streams.

That trees are great promoters of lakes and rivers appears
from a well-known fact in North America; for, since the woods
and forests have been grubbed and cleared, all bodies of water
are much diminished; so that some streams, that were very
considerable a century ago, will not now drive a common mill.*
Besides, most woodlands, forests, and chases with us abound
with pools and morasses; no doubt for the reason given above.

To a thinking mind few phenomena are more strange than
the state of little ponds on the summits of chalk-hills, many of
which are never dry in the most trying droughts of summer.
On chalk-hills I say, because in many rocky and gravelly soils
springs usually break out pretty high on the sides of elevated
grounds and mountains; but no person acquainted with chalky
districts will allow that they ever saw springs in such a soil but
in valleys and bottoms, since the waters of so pervious a stratum
as chalk all lie on one dead level, as well-diggers have assured
me again and again.

Now we have many such little round ponds in this district;
and one in particular on our sheep-down, three hundred feet
above my house; which though never above three feet deep in
the middle, and not more than thirty feet in diameter, and
containing perhaps not more than two or three hundred hogsheads of water, yet never is known to fail, though it affords
drink for three hundred or four hundred sheep, and for at least
twenty head of large cattle beside. This pond, it is true, is over-
hung with two moderate beeches, that, doubtless, at times
afford it much supply: but then we have others as small, that,
without the aid of trees, and in spite of evaporation from sun
and wind, and perpetual consumption by cattle, yet constantly
maintain a moderate share of water, without overflowing in the
wettest of seasons, as they would do if supplied by springs. By
my journal of May 1775, it appears that 'the small and even

*Vide Kalm's Travels to North America.

considerable ponds in the vales are now dried up, while the small ponds on the very tops of hills are but little affected.' Can this difference be accounted for from evaporation alone, which certainly is more prevalent in bottoms? or rather have not those elevated pools some unnoticed recruits, which in the night time counterbalance the waste of the day; without which the cattle alone must soon exhaust them? And here it will be necessary to enter more minutely into the cause. Dr Hales,[145] in his *Vegetable Statics*, advances, from experiment, that 'the moister the earth is the more dew falls on it in a night: and more than a double quantity of dew falls on a surface of water than there does on an equal surface of moist earth.' Hence we see that water, by its coolness, is enabled to assimilate to itself a large quantity of moisture nightly by condensation;[146] and that the air, when loaded with fogs and vapours, and even with copious dews, can alone advance a considerable and never-failing resource. Persons that are much abroad, and travel early and late, such as shepherds, fishermen, etc., can tell what prodigious fogs prevail in the night on elevated downs, even in the hottest parts of summer; and how much the surfaces of things are drenched by those swimming vapours, though, to the senses, all the while, little moisture seems to fall.

I am, etc.

LETTER XXX

Selborne, April 3, 1776.

DEAR SIR,

Monsieur Herissant, a French anatomist, seems persuaded that he has discovered the reason why cuckoos do not hatch their own eggs; the impediment, he supposes, arises from the internal structure of their parts, which incapacitates them for incubation. According to this gentleman, the crop or craw of a cuckoo does not lie before the *sternum* at the bottom of the neck, as in the *gallinæ*, *columbæ*, etc., but immediately behind it, on and over the bowels, so as to make a large protuberance in the belly.*

Induced by this assertion, we procured a cuckoo; and, cutting open the breast-bone, and exposing the intestines to sight, found the crop lying as mentioned above. This stomach was large and round, and stuffed hard like a pin-cushion with food, which, upon nice examination, we found to consist of various insects; such as small scarabs, spiders, and dragon-flies; the last of which we have seen cuckoos catching on the wing as they were just emerging out of the *aurelia* state. Among this farrago also were to be seen maggots, and many seeds, which belonged either to gooseberries, currants, cranberries, or some such fruit; so that these birds apparently subsist on insects and fruits: nor was there the least appearance of bones, feathers, or fur to support the idle notion of their being birds of prey.

The *sternum* in this bird seemed to us to be remarkably short, between which and the *anus* lay the crop, or craw, and immediately behind that the bowels against the backbone.

It must be allowed, as this anatomist observes, that the crop placed just upon the bowels must, especially when full, be in a very uneasy situation during the business of incubation; yet the test will be to examine whether birds that are actually known to sit for certain are not formed in a similar manner. This inquiry I proposed to myself to make with a fern-owl, or goat-sucker, as soon as opportunity offered: because, if their information proves the same, the reason for incapacity in the cuckoo will be allowed to have been taken up somewhat hastily.

Not long after a fern-owl was procured, which, from its habit and shape, we suspected might resemble the cuckoo in its internal construction. Nor were our suspicions ill-grounded; for, upon the dissection, the crop, or craw, also lay behind the *sternum*, immediately on the viscera, between them and the

* *Histoire de l'Academie Royale*, 1752.

skin of the belly. It was bulky, and stuffed hard with large *phalænæ*, moths of several sorts, and their eggs, which no doubt had been forced out of those insects by the action of swallowing.

Now as it appears that this bird, which is so well known to practise incubation, is formed in a similar manner with cuckoos, Monsieur Herissant's conjecture, that cuckoos are incapable of incubation from the disposition of their intestines, seems to fall to the ground: and we are still at a loss for the cause of that strange and singular peculiarity in the instance of the *cuculus canorus*.

We found the case to be the same with the ring-tail hawk, in respect to formation; and, as far as I can recollect, with the swift; and probably it is so with many more sorts of birds that are not granivorous.

I am, etc.

LETTER XXXI

Selborne, April 29, 1776.

DEAR SIR,

On August the 4th, 1775, we surprised a large viper, which seemed very heavy and bloated, as it lay in the grass basking in the sun. When we came to cut it up, we found that the abdomen was crowded with young, fifteen in number; the shortest of which measured full seven inches, and were about the size of full-grown earthworms. This little fry issued into the world with the true viper-spirit about them, showing great alertness as soon as disengaged from the belly of the dam: they twisted and wriggled about, and set themselves up, and gaped very wide when touched with a stick, showing manifest tokens of menace and defiance, though as yet they had no manner of fangs that we could find, even with the help of our glasses.

To a thinking mind nothing is more wonderful than that early instinct which impresses young animals with the notion of the situation of their natural weapons, and of using them properly in their own defence, even before those weapons subsist or are formed. Thus a young cock will spar at his adversary before his spurs are grown; and a calf or a lamb will push with their heads before their horns are sprouted. In the same manner did these young adders attempt to bite before their fangs were in being. The dam however was furnished with very formidable ones, which we lifted up (for they fold down when not used) and cut them off with the point of our scissors.

There was little room to suppose that this brood had ever been in the open air before; and that they were taken in for refuge, at the mouth of the dam, when she perceived that danger was approaching; because then probably we should have found them somewhere in the neck, and not in the abdomen.

LETTER XXXII

CASTRATION has a strange effect: it emasculates both man, beast, and bird, and brings them to a near resemblance of the other sex. Thus eunuchs have smooth unmuscular arms, thighs, and legs; and broad hips, and beardless chins, and squeaking voices. Gelt-stags and bucks have hornless heads, like hinds and does. Thus wethers have small horns, like ewes; and oxen large bent horns, and hoarse voices when they low, like cows: for bulls have short straight horns; and though they mutter and grumble in a deep tremendous tone, yet they low in a shrill high key. Capons have small combs and gills, and look pallid about the head, like pullets; they also walk without any parade, and hover chickens like hens. Barrow-hogs have also small tusks like sows.

Thus far it is plain that the deprivation of masculine vigour puts a stop to the growth of those parts or appendages that are looked upon as its insignia. But the ingenious Mr Lisle,[147] in his book on husbandry, carries it much farther; for he says that the loss of those insignia alone has sometimes a strange effect on the ability itself: he had a boar so fierce and venereous, that, to prevent mischief, orders were given for his tusks to be broken off. No sooner had the beast suffered this injury than his powers forsook him, and he neglected those females to whom before he was passionately attached, and from whom no fences could restrain him.

LETTER XXXIII

THE natural term of an hog's life is little known, and the reason is plain – because it is neither profitable nor convenient to keep that turbulent animal to the full extent of its time: however, my neighbour, a man of substance, who had no occasion to study every little advantage to a nicety, kept an half-bred Bantam sow, who was as thick as she was long, and whose belly swept on the ground, till she was advanced to her seventeenth year; at which period she showed some tokens of age by the decay of her teeth and the decline of her fertility.

For about ten years this prolific mother produced two litters in the year of about ten at a time, and once above twenty at a litter; but, as there were near double the number of pigs to that of teats, many died. From long experience in the world this female was grown very sagacious and artful: when she found occasion to converse with a boar she used to open all the intervening gates, and march, by herself, up to a distant farm where one was kept; and when her purpose was served would return by the same means. At the age of about fifteen her litters began to be reduced to four or five; and such a litter she exhibited when in her fatting-pen. She proved, when fat, good bacon, juicy, and tender; the rind, or sward, was remarkably thin. At a moderate computation she was allowed to have been the fruitful parent of three hundred pigs: a prodigious instance of fecundity in so large a quadruped! She was killed in spring 1775.

I am, etc.

LETTER XXXIV

Selborne, May 9, 1776.

DEAR SIR,

... admôrunt ubera tigres.

We have remarked in a former letter how much incongruous animals, in a lonely state, may be attached to each other from a spirit of sociality; in this it may not be amiss to recount a different motive which has been known to create as strange a fondness.

My friend had a little helpless leveret brought to him, which the servants fed with milk in a spoon, and about the same time his cat kittened and the young were dispatched and buried. The hare was soon lost, and supposed to be gone the way of most foundlings, to be killed by some dog or cat. However, in about a fortnight, as the master was sitting in his garden in the dusk of the evening, he observed his cat, with tail erect, trotting towards him, and calling with little short inward notes of complacency, such as they use towards their kittens, and something gamboling after, which proved to be the leveret that the cat had supported with her milk, and continued to support with great affection.

Thus was a graminivorous animal nurtured by a carnivorous and predaceous one!

Why so cruel and sanguinary a beast as a cat, of the ferocious genus of *Feles*, the *murium leo*,[148] as Linnæus calls it, should be affected with any tenderness towards an animal which is its natural prey, is not so easy to determine.

This strange affection probably was occasioned by that *desiderium*, those tender maternal feelings, which the loss of her kittens had awakened in her breast; and by the complacency and ease she derived to herself from the procuring her teats to be drawn, which were too much distended with milk, till, from habit, she became as much delighted with this foundling as if it had been her real offspring.

This incident is no bad solution of that strange circumstance which grave historians as well as the poets assert, of exposed children being sometimes nurtured by female wild beasts that probably had lost their young. For it is not one whit more marvellous that Romulus and Remus, in their infant state, should be nursed by a she-wolf, than that a poor little sucking leveret should be fostered and cherished by a bloody grimalkin.

> . . . viridi fœtam Mavortis in antro
> Procubuisse lupam: geminos huic ubera circum
> Ludere pendentes pueros, et lambere matrem
> Impavidos: illam tereti cervice reflexam
> Mulcere alternos, et corpora fingere linguâ.

LETTER XXXV

Selborne, May 20, 1777.

DEAR SIR,

Lands that are subject to frequent inundations are always poor; and probably the reason may be because the worms are drowned. The most insignificant insects and reptiles are of much more consequence, and have much more influence in the œconomy of nature, than the incurious are aware of; and are mighty in their effect, from their minuteness, which renders them less an object of attention; and from their numbers and fecundity. Earth-worms, though in appearance a small and despicable link in the chain of nature, yet, if lost, would make a lamentable chasm. For, to say nothing of half the birds, and some quadrupeds, which are almost entirely supported by them, worms seem to be the great promoters of vegetation, which would proceed but lamely without them, by boring, perforating, and loosening the soil, and rendering it pervious to rains and the fibres of plants, by drawing straws and stalks of leaves and twigs into it; and, most of all, by throwing up such infinite numbers of lumps of earth called worm-casts, which, being their excrement, is a fine manure for grain and grass. Worms probably provide new soil for hills and slopes where the rain washes the earth away; and they affect slopes, probably to avoid being flooded. Gardeners and farmers express their detestation of worms; the former because they render their walks unsightly, and make them much work: and the latter because, as they think, worms eat their green corn. But these men would find that the earth without worms would soon become cold, hard-bound, and void of fermentation; and consequently sterile: and besides, in favour of worms, it should be hinted that green corn, plants, and flowers,are not so much injured by them as by many species of *coleoptera* (scarabs), and *tipulæ* (long-legs), in their larva, or grub-state; and by unnoticed myriads of small shell-less snails, called slugs, which silently and imperceptibly make amazing havoc in the field and garden.*

These hints we think proper to throw out in order to set the inquisitive and discerning to work.

A good monography of worms[149] would afford much entertainment and information at the same time, and would open a large and new field in natural history. Worms work most in the spring; but by no means lie torpid in the dead months; are out every mild night in the winter, as any person may be convinced that will take the pains to examine his grass-plots with a candle; are hermaphrodites, and much addicted to venery, and consequently very prolific.

I am, etc.

LETTER XXXVI

Selborne, Nov. 22, 1777.

Dear Sir,

You cannot but remember that the twenty-sixth and twenty-seventh of last March were very hot days; so sultry that everybody complained and were restless under those sensations to which they had not been reconciled by gradual approaches.

This sudden summer-like heat was attended by many summer coincidences; for on those two days the thermometer rose to sixty-six in the shade; many species of insects revived and came forth; some bees swarmed in this neighbourhood; the old tortoise, near Lewes in Sussex, awakened and came forth out of its dormitory; and, what is most to my present

---

*Farmer Young, of Norton-farm, says that this spring (1777) about four acres of his wheat in one field was entirely destroyed by slugs, which swarmed on the blades of corn, and devoured it as fast as it sprang.

purpose, many house-swallows appeared and were very alert in many places, and particularly at Cobham, in Surrey.

But as that short warm period was succeeded as well as preceded by harsh severe weather, with frequent frosts and ice, and cutting winds, the insects withdrew, the tortoise retired again into the ground, and the swallows were seen no more until the tenth of April, when, the rigour of the spring abating, a softer season began to prevail.

Again; it appears by my journals for many years past, that house-martins retire, to a bird, about the beginning of October; so that a person not very observant of such matters would conclude that they had taken their last farewell: but then it may be seen in my diaries also that considerable flocks have discovered themselves again in the first week of November, and often on the fourth day of that month only for one day; and that not as if they were in actual migration, but playing about at their leisure and feeding calmly, as if no enterprise of moment at all agitated their spirits. And this was the case in the beginning of this very month; for, on the fourth of November, more than twenty house-martins, which, in appearance, had all departed about the seventh of October, were seen again, for that one morning only, sporting between my fields and the Hanger, and feasting on insects which swarmed in that sheltered district. The preceding day was wet and blustering, but the fourth was dark and mild, and soft, the wind at south-west, and the thermometer at $58\frac{1}{2}°$; a pitch not common at that season of the year. Moreover, it may not be amiss to add in this place, that whenever the thermometer is above 50° the bat comes flitting out in every autumnal and winter month.

From all these circumstances laid together, it is obvious that torpid insects, reptiles, and quadrupeds, are awakened from their profoundest slumbers by a little untimely warmth; and therefore that nothing so much promotes this death-like stupor as a defect of heat. And farther, it is reasonable to suppose that two whole species, or at least many individuals of those two species, of British *hirundines*, do never leave this island at all, but partake of the same benumbed state: for we cannot suppose that, after a month's absence, house-martins can return from southern regions to appear for one morning in November, or that house-swallows should leave the districts of Africa to enjoy, in March, the transient summer of a couple of days.

I am, etc.

LETTER XXXVII

Selborne, Jan. 8, 1778.

DEAR SIR,

There was in this village several years ago a miserable pauper, who, from his birth, was afflicted with a leprosy, as far as we are aware of a singular kind, since it affected only the palms of his hands and the soles of his feet. This scaly eruption usually broke out twice in the year, at the spring and fall; and, by peeling away, left the skin so thin and tender that neither his hands or feet were able to perform their functions; so that the poor object was half his time on crutches, incapable of employ, and languishing in a tiresome state of indolence and inactivity. His habit was lean, lank, and cadaverous. In this sad plight he dragged on a miserable existence, a burden to himself and his parish, which was obliged to support him till he was relieved by death at more than thirty years of age.

The good women, who love to account for every defect in children by the doctrine of longing, said that his mother felt a violent propensity for oysters, which she was unable to gratify; and that the black rough scurf on his hands and feet were the shells of that fish. We knew his parents, neither of which were lepers; his father in particular lived to be far advanced in years.

In all ages the leprosy has made dreadful havoc among mankind. The Israelites seem to have been greatly afflicted with it from the most remote times; as appears from the peculiar and repeated injunctions given them in the Levitical law.* Nor was the rancour of this foul disorder much abated in the last period of their commonwealth, as may be seen in many passages of the New Testament.

Some centuries ago this horrible distemper prevailed all Europe over; and our forefathers were by no means exempt, as appears by the large provision made for objects labouring under this calamity. There was an hospital for female lepers in the diocese of Lincoln, a noble one near Durham, three in London and Southwark, and perhaps many more in or near our great towns and cities. Moreover, some crowned heads, and other wealthy and charitable personages, bequeathed large legacies to such poor people as languished under this hopeless infirmity.

It must therefore, in these days, be, to an humane and thinking person, a matter of equal wonder and satisfaction, when he contemplates how nearly this pest is eradicated, and observes that a leper now is a rare sight. He will, moreover, when engaged in such a train of thought, naturally inquire for the reason. This happy change perhaps may have originated and been continued from the much smaller quantity of salted meat and fish now eaten in these kingdoms; from the use of linen next the skin; from the plenty of better bread; and from the profusion of fruits, roots, legumes, and greens, so common in every family. Three or four centuries ago, before there were any enclosures, sown-grasses, field-turnips, or field-carrots, or hay, all the cattle which had grown fat in summer, and were not killed for winter-use, were turned out soon after Michaelmas to shift as they could through the dead months; so that no fresh meat could be had in winter or spring. Hence the marvellous account of the vast stores of salted flesh found in the larder of the eldest Spencer† in the days of Edward the Second, even so late in the spring as the third of May. It was from magazines

* See Leviticus, chap. xiii. and xiv.

† *Viz.*: Six hundred bacons, eighty carcasses of beef, and six hundred muttons.

like these that the turbulent barons supported in idleness their
riotous swarms of retainers ready for any disorder or mischief.
But agriculture is now arrived at such a pitch of perfection, that
our best and fattest meats are killed in the winter; and no man
need eat salted flesh, unless he prefers it, that has money to buy
fresh.

One cause of this distemper might be, no doubt, the quantity
of wretched fresh and salt fish consumed by the commonalty at
all seasons as well as in Lent; which our poor now would hardly
be persuaded to touch.

The use of linen changes, shirts or shifts, in the room of sordid
and filthy woollen, long worn next the skin, is a matter of
neatness comparatively modern; but must prove a great means
of preventing cutaneous ails. At this very time woollen instead
of linen prevails among the poorer Welch, who are subject to
foul eruptions.

The plenty of good wheaten bread that now is found among
all ranks of people in the south, instead of that miserable sort
which used in old days to be made of barley or beans, may
contribute not a little to the sweetening their blood and correc-
ting their juices; for the inhabitants of mountainous districts,
to this day, are still liable to the itch and other cutaneous dis-
orders, from a wretchedness and poverty of diet.

As to the produce of a garden, every middle-aged person of
observation may perceive, within his own memory, both in
town and country, how vastly the consumption of vegetables
is increased. Green-stalls in cities now support multitudes in a
comfortable state, while gardeners get fortunes. Every decent
labourer also has his garden, which is half his support, as well as
his delight; and common farmers provide plenty of beans, peas,
and greens, for their hinds[150] to eat with their bacon; and those
few that do not are despised for their sordid parsimony, and
looked upon as regardless of the welfare of their dependents.
Potatoes have prevailed in this little district, by means of
premiums, within these twenty years only; and are much

esteemed here now by the poor, who would scarce have ventured to taste them in the last reign.

Our Saxon ancestors certainly had some sort of cabbage, because they call the month of February sprout-cale; but, long after their days, the cultivation of gardens was little attended to. The religious, being men of leisure, and keeping up a constant correspondence with Italy, were the first people among us that had gardens and fruit-trees in any perfection, within the walls of their abbies* and priories. The barons neglected every pursuit that did not lead to war or tend to the pleasure of the chase.

It was not till gentlemen took up the study of horticulture themselves that the knowledge of gardening made such hasty advances. Lord Cobham, Lord Ila, and Mr Waller of Beaconsfield, were some of the first people of rank that promoted the elegant science of ornamenting without despising the superintendence of the kitchen quarters and fruit walls.

A remark made by the excellent Mr Ray in his Tour of Europe at once surprises us, and corroborates what has been advanced above; for we find him observing, so late as his days, that 'the Italians use several herbs for sallets, which are not yet or have not been but lately used in England, viz., *selleri* (celery), which is nothing else but the sweet smallage; the young shoots whereof, with a little of the head of the root cut off, they eat raw with oil and pepper.' And further he adds 'curled endive blanched is much used beyond seas; and, for a raw sallet, seemed to excel lettuce itself.' Now this journey was undertaken no longer ago than in the year 1663.

I am, etc.

---

* 'In monasteries the lamp of knowledge continued to burn, however dimly. In them men of business were formed for the state: the art of writing was cultivated by the monks; they were the only proficients in mechanics, gardening, and architecture.' – See Dalrymple's *Annals of Scotland.*

LETTER XXXVIII

Fortè puer, comitum seductus ab agmine fido,
Dixerat, ecquis adest? et, adest, responderat echo.
Hic stupet; utque aciem partes divisit in omnes;
Voce, veni, clamat magnâ. Vocat illa vocantem.

Selborne, Feb. 12, 1778.

DEAR SIR,

In a district so diversified as this, so full of hollow vales, and hanging woods, it is no wonder that echoes should abound. Many we have discovered that return the cry of a pack of dogs, the notes of a hunting-horn, a tunable ring of bells, or the melody of birds, very agreeably: but we were still at a loss for a polysyllabical, articulate echo, till a young gentleman, who had parted from his company in a summer evening walk, and was calling after them, stumbled upon a very curious one in a spot where it might least be expected. At first he was much surprised, and could not be persuaded but that he was mocked by some boy; but, repeating his trials in several languages, and finding his respondent to be a very adroit polyglot, he then discerned the deception.

This echo in an evening, before rural noises cease, would repeat ten syllables most articulately and distinctly, especially if quick dactyls were chosen. The last syllables of

Tityre, tu patulæ recubans . . .

were as audibly and intelligibly returned as the first: and there is no doubt, could trial have been made, but that at midnight, when the air is very elastic, and a dead stillness prevails, one or two syllables more might have been obtained; but the distance rendered so late an experiment very inconvenient.

Quick dactyls, we observed, succeeded best; for when we came to try its powers in slow, heavy, embarrassed spondees of the same number of syllables,

Monstrum horrendum, informe, ingens . . .

we could perceive a return but of four or five.

All echoes have some one place to which they are returned stronger and more distinct than to any other; and that is always the place that lies at right angles with the object of repercussion, and is not too near, nor too far off. Buildings, or naked rocks, re-echo much more articulately than hanging wood or vales; because in the latter the voice is as it were entangled, and embarrassed in the covert, and weakened in the rebound.

The true object of this echo, as we found by various experiments, is the stone-built, tiled hop-kiln in Galley-lane, which measures in front 40 feet, and from the ground to the eaves 12 feet. The true *centrum phonicum*, or just distance, is one particular spot in the King's-field, in the path to Nore-hill, on the very brink of the steep balk above the hollow cart way. In this case there is no choice of distance; but the path, by mere contingency, happens to be the lucky, the identical spot, because the ground rises or falls so immediately, if the speaker either retires or advances, that his mouth would at once be above or below the object.

We measured this polysyllabical echo with great exactness, and found the distance to fall very short of Dr Plot's rule for distinct articulation: for the Doctor, in his history of Oxfordshire, allows 120 feet for the return of each syllable distinctly: hence this echo, which gives ten distinct syllables, ought to measure 400 yards, or 120 feet to each syllable; whereas our distance is only 258 yards, or near 75 feet, to each syllable. Thus our measure falls short of the Doctor's, as five to eight: but then it must be acknowledged that this candid philosopher was convinced afterwards, that some latitude must be admitted of in the distance of echoes according to time and place.

When experiments of this sort are making, it should always be remembered that weather and the time of day have a vast influence on an echo; for a dull, heavy, moist air deadens and clogs the sound; and hot sunshine renders the air thin and weak, and deprives it of all its springiness; and a ruffling wind quite defeats the whole. In a still, clear, dewy evening the air is most elastic; and perhaps the later the hour the more so.

Echo has always been so amusing to the imagination, that the poets have personified her; and in their hands she has been the occasion of many a beautiful fiction. Nor need the gravest man be ashamed to appear taken with such a phenomenon, since it may become the subject of philosophical or mathematical inquiries.

One should have imagined that echoes, if not entertaining, must at least have been harmless and inoffensive; yet Virgil advances a strange notion, that they are injurious to bees. After enumerating some probable and reasonable annoyances, such as prudent owners would wish far removed from their bee-gardens, he adds

... aut ubi concava pulsu
Saxa sonant, vocisque offensa resultat imago.

This wild and fanciful assertion will hardly be admitted by the philosophers of these days; especially as they all now seem agreed that insects are not furnished with any organs of hearing at all.[151] But if it should be urged, that though they cannot hear yet perhaps they may feel the repercussion of sounds, I grant it is possible they may. Yet that these impressions are distasteful or hurtful, I deny, because bees, in good summers, thrive well in my outlet, where the echoes are very strong: for this village is another Anathoth, a place of responses or echoes. Besides, it does not appear from experiment that bees are in any way capable of being affected by sounds: for I have often tried my own with a large speaking-trumpet held close to their hives, and with such an exertion of voice as would have hailed a ship at the distance of a mile, and still these insects pursued their various employments undisturbed, and without showing the least sensibility or resentment.

Some time since its discovery this echo is become totally

silent, though the object, or hop-kiln remains: nor is there any mystery in this defect, for the field between is planted as an hop-garden, and the voice of the speaker is totally absorbed and lost among the poles and entangled foliage of the hops. And when the poles are removed in autumn the disappointment is the same; because a tall quick-set hedge, nurtured up for the purpose of shelter to the hop ground, entirely interrupts the impulse and repercussion of the voice: so that till those obstructions are removed no more of its garrulity can be expected.

Should any gentleman of fortune think an echo in his park or outlet a pleasing incident, he might build one at little or no expense. For whenever he had occasion for a new barn, stable, dog-kennel, or the like structure, it would be only needful to erect this building on the gentle declivity of an hill, with a like rising opposite to it, at a few hundred yards distance; and perhaps success might be the easier ensured could some canal, lake, or stream, intervene. From a seat at the *centrum phonicum* he and his friends might amuse themselves sometimes of an evening with the prattle of this loquacious nymph; of whose complacency and decent reserve more may be said than can with truth of every individual of her sex; since she is

> ... quæ nec reticere loquenti,
> Nec prior ipsa loqui didicit resonabilis echo.

I am, etc.

P.S. – The classic reader will, I trust, pardon the following lovely quotation, so finely describing echoes, and so poetically accounting for their causes from popular superstition:

> Quæ benè quom videas, rationem reddere possis
> Tute tibi atque aliis, quo pacto per loca sola
> Saxa pareis formas verborum ex ordine reddant,
> Palanteis comites quom monteis inter opacos
> Quærimus, et magna dispersos voce ciemus.
> Sex etiam, aut septem loca vidi reddere voces
> Unam quom jaceres: ita colles collibus ipsis

Verba repulsantes iterabant dicta referre.
Hæc loca capripedes Satyros, Nymphasque tenere
Finitimi fingunt, et Faunos esse loquuntur;
Quorum noctivago strepitu, ludoque jocanti
Adfirmant volgo taciturna silentia rumpi,
Chordarumque sonos fieri, dulceisque querelas,
Tibia quas fundit digitis pulsata canentum:
Et genus agricolûm latè sentiscere, quom Pan
Pinea semiferi capitis velamina quassans,
Unco sæpe labro calamos percurrit hianteis,
Fistula silvestrem ne cesset fundere musam.

LUCRETIUS, lib. iv. l. 576.

LETTER XXXIX

Selborne, May 13, 1778.

DEAR SIR,

Among the many singularities attending those amusing birds the swifts, I am now confirmed in the opinion that we have every year the same number of pairs invariably; at least the result of my inquiry has been exactly the same for a long time past. The swallows and martins are so numerous, and so widely distributed over the village, that it is hardly possible to recount them; while the swifts, though they do not all build in the church, yet so frequently haunt it, and play and rendezvous round it, that they are easily enumerated. The number that I constantly find are eight pairs; about half of which reside in the church, and the rest build in some of the lowest and meanest thatched cottages. Now as these eight pairs, allowance being made for accidents, breed yearly eight pairs more, what becomes annually of this increase; and what determines every spring which pairs shall visit us, and reoccupy their ancient haunts?

Ever since I have attended to the subject of ornithology, I have always supposed that that sudden reverse of affection, that strange ἀντιστοργή, which immediately succeeds in the feathered kind to the most passionate fondness, is the occasion of an equal dispersion of birds over the face of the earth. Without this provision one favourite district would be crowded with inhabitants, while others would be destitute and forsaken. But the parent birds seem to maintain a jealous superiority, and to oblige the young to seek for new abodes:[152] and the rivalry of the males, in many kinds, prevents their crowding the one on the other. Whether the swallows and house-martins return in the same exact number annually is not easy to say, for reasons given above: but it is apparent, as I have remarked before in my Monographies, that the numbers returning bear no manner of proportion to the numbers retiring.

LETTER XL

Selborne, June 2, 1778.

DEAR SIR,

The standing objection to botany has always been, that it is a pursuit that amuses the fancy and exercises the memory, without improving the mind or advancing any real knowledge: and where the science is carried no farther than a mere systematic classification, the charge is but too true. But the botanist that is desirous of wiping off this aspersion should be by no means content with a list of names; he should study plants philosophically, should investigate the laws of vegetation, should examine the powers and virtues of efficacious herbs, should promote their cultivation; and graft the gardener, the planter, and the husbandman, on the phytologist. Not that system is by any means to be thrown aside; without system the field of nature would be a pathless wilderness: but system should be subservient to, not the main object of, pursuit.

Vegetation is highly worthy of our attention; and in itself is of the utmost consequence to mankind, and productive of many of the greatest comforts and elegancies of life. To plants we owe timber, bread, beer, honey, wine, oil, linen, cotton, etc., what not only strengthens our hearts, and exhilarates our spirits, but what secures from inclemencies of weather and adorns our persons. Man, in his true state of nature, seems to be subsisted by spontaneous vegetation: in middle climes, where grasses prevail, he mixes some animal food with the produce of the field and garden: and it is towards the polar extremes only that, like his kindred bears and wolves, he gorges himself with flesh alone, and is driven, to what hunger has never been known to compel the very beasts, to prey on his own species.*

The productions of vegetation have had a vast influence on the commerce of nations, and have been the great promoters of navigation, as may be seen in the articles of sugar, tea, tobacco, opium, ginseng, betel, paper, etc. As every climate has its peculiar produce, our natural wants bring on a mutual intercourse; so that by means of trade each distant part is supplied with the growth of every latitude. But, without the knowledge of plants and their culture, we must have been content with our hips and haws, without enjoying the delicate fruits of India and the salutiferous drugs of Peru.

Instead of examining the minute distinctions of every various species of each obscure genus, the botanist should endeavour to make himself acquainted with those that are useful. You shall see a man readily ascertain every herb of the field, yet hardly know wheat from barley, or at least one sort of wheat or barley from another.

But of all sorts of vegetation the grasses seem to be most neglected; neither the farmer nor the grazier seem to distinguish the annual from the perennial, the hardy from the tender, nor the succulent and nutritive from the dry and juiceless.

The study of grasses would be of great consequence to a

* See the late *Voyages to the South-seas*.

northerly and grazing kingdom. The botanist that could improve the sward of the district where he lived would be an useful member of society; to raise a thick turf on a naked soil would be worth volumes of systematic knowledge; and he would be the best commonwealth's man that could occasion the growth of 'two blades of grass where one alone was seen before.'

I am, etc.

LETTER XLI

Selborne, July 3, 1778.

DEAR SIR,

In a district so diversified with such a variety of hill and dale, aspects, and soils, it is no wonder that great choice of plants should be found. Chalks, clays, sands, sheep-walks and downs, bogs, heaths, woodlands and champaign fields,[153] cannot but furnish an ample flora. The deep rocky lanes abound with *filices*, and the pastures and moist woods with *fungi*. If in any branch of botany we may seem to be wanting, it must be in the large aquatic plants, which are not to be expected on a spot far removed from rivers, and lying up amidst the hill country at the spring heads. To enumerate all the plants that have been discovered within our limits would be a needless work; but a short list of the more rare, and the spots where they are to be found, may be neither unacceptable nor unentertaining:

Helleborus fœtidus, stinking hellebore, bear's foot, or setterworth, – all over the High-wood and Coney-crofthanger: this continues a great branching plant the winter through, blossoming about January, and is very ornamental in shady walks and shrubberies. The good women give the leaves powdered to children troubled with worms; but it is a violent remedy, and ought to be administered with caution.

Helleborus viridis, green hellebore, – in the deep stony lane on the left hand just before the turning to Nortonfarm, and at the top of Middle Dorton under the hedge: this plant dies down to the ground early in autumn, and springs again about February, flowering almost as soon as it appears above ground.

Vaccinium oxycoccos, creeping bilberries or cranberries, – in the bogs of Bin's-pond.

Vaccinium myrtillus, whortle, or bleaberries, – on the dry hillocks of Wolmer-forest.

Drosera rotundifolia, round-leaved sun-⎫ In the bogs
dew. *Drosera longifolia*, long-leaved ditto. ⎭ of Bin's-pond.

Comarum palustre, purple comarum, or marsh cinquefoil, – in the bogs of Bin's-pond.

Hypericon androsæmum, Tutsan, St John's wort, – in the stony, hollow lanes.

Vinca minor, less periwinkle, – in Selborne Hanger and Shrubwood.

Monotropa hypopithys, yellow monotropa, or bird's nest, – in Selborne Hanger under the shady beeches, to whose roots it seems to be parasitical – at the north-west end of the Hanger.

Chlora perfoliata, Blackstonia perfoliata, Hudsoni, perfoliated yellow-wort, – on the banks in the King's-field.

Paris quadrifolia, herb Paris, true-love, or one-berry, – in the Church Litten coppice.

Chrysosplenium oppositifolium, opposite golden saxifrage, – in the dark and rocky hollow lanes.

Gentiana amarella, autumnal gentian or fellwort, – on the Zig-zag and Hanger;

Lathræa squamaria, tooth-wort, – in the Church Litten coppice under some hazels near the foot-bridge, in Trimming's garden-hedge, and on the dry wall opposite Grange-yard.

Dipsacus pilosus, small teasel, – in the Short and Long Lith.

Lathyrus sylvestris, narrow-leaved, or wild lathyrus,[154] – in the bushes at the foot of the Short Lith, near the path.

Ophrys spiralis, ladies' traces, – in the Long Lith, and towards the south-corner of the common.

Ophrys nidus avis, birds' nest ophrys,[155] – in the Long Lith under the shady beeches among the dead leaves; in Great Dorton among the bushes, and on the Hanger plentifully.

Serapias latifolia, helleborine,[156] – in the High-wood under the shady beeches.

Daphne laureola, spurge laurel, – in Selborne Hanger and the High-wood.

Daphne mezereum, the mezereon, – in Selborne Hanger among the shrubs at the south-east end above the cottages.

Lycoperdon tuber, truffles, – in the Hanger and High-wood.

Sambucus ebulus, dwarf elder, walwort, or danewort, – among the rubbish and ruined foundations of the Priory.

Of all the propensities of plants[157] none seem more strange than their different periods of blossoming. Some produce their flowers in the winter, or very first dawnings of spring; many when the spring is established; some at midsummer, and some not till autumn. When we see the *helleborus fœtidus* and *helleborus niger* blowing at Christmas, the *helleborus hyemalis* in January, and the *helleborus viridis* as soon as ever it emerges out of the ground, we do not wonder, because they are kindred plants that we expect should keep pace the one with the other. But other congenerous vegetables differ so widely in their time of flowering that we cannot but admire. I shall only instance at present in the *crocus sativus*, the vernal, and the autumnal crocus, which have such an affinity, that the best botanists only make them varieties of the same genus, of which there is only one species;[158] not being able to discern any difference in the corolla, or in the internal structure. Yet the vernal crocus expands its flowers by the beginning of March at farthest, and often in very rigorous weather; and cannot be retarded but by some violence offered: while the autumnal (the saffron) defies the influence of the spring and summer, and will not blow till most plants begin to fade and run to seed. This circumstance is one of the wonders of the creation, little noticed, because a common occurrence: yet ought not to be overlooked on account of its being familiar, since it would be as difficult to be explained as the most stupendous phænomenon in nature.

> Say, what impels, amidst surrounding snow,
> Congealed, the crocus' flamy bud to grow?
> Say, what retards, amidst the summer's blaze,
> Th' autumnal bulb till pale, declining days?
> The GOD of SEASONS; whose pervading power
> Controls the sun, or sheds the fleecy shower:
> He bids each flower His quickening word obey;
> Or to each lingering bloom enjoins delay.

LETTER XLII

Omnibus animalibus reliquis certus et uniusmodi, et in suo cuique genere incessus est: aves solæ vario meatu feruntur, et in terrâ, et in aëre. [159]

PLIN. *Hist. Nat.* lib. x. cap. 38.

Selborne, Aug. 7, 1778.

DEAR SIR,

A good ornithologist should be able to distinguish birds by their air as well as by their colours and shape; on the ground as well as on the wind, and in the bush as well as in the hand. For, though it must not be said that every species of birds has a manner peculiar to itself, yet there is somewhat in most genera at least, that at first sight discriminates them, and enables a judicious observer to pronounce upon them with some certainty. Put a bird in motion.

... Et verâ incessu patuit. ...

Thus kites and buzzards sail round in circles with wings expanded and motionless; and it is from their gliding manner that the former are still called in the north of England gleads, from the Saxon verb *glidan* to glide. The kestrel, or wind-hover, has a peculiar mode of hanging in the air in one place, his wings all the while being briskly agitated. Hen-carriers fly low over heaths or fields of corn, and beat the ground regularly like a pointer or setting-dog. Owls move in a buoyant manner, as if lighter than the air; they seem to want ballast. There is a peculiarity belonging to ravens that must draw the attention even of the most incurious – they spend all their leisure time in striking and cuffing each other on the wing in a kind of playful skirmish; and, when they move from one place to another, frequently turn on their backs with a loud croak, and seem to be falling to the ground. When this odd gesture betides them, they are scratching themselves with one foot,[160] and thus lose the centre of gravity. Rooks sometimes dive and tumble in a frolicsome manner; crows and daws swagger in their walk; wood-peckers fly *volatu undoso*, opening and closing their wings at every stroke, and so are always rising or falling in curves. All of this genus use their tails, which incline downward, as a support while they run up trees. Parrots, like all other hook-clawed birds, walk awkwardly, and make use of their bill as a third foot, climbing and ascending with ridiculous caution. All the *gallinæ* parade and walk gracefully, and run nimbly; but fly with difficulty, with an impetuous whirring, and in a straight line. Magpies and jays flutter with powerless wings, and make no dispatch; herons seem incumbered with too much sail for their light bodies; but these vast hollow wings are necessary in carrying burdens, such as large fishes, and the like; pigeons, and particularly the sort called smiters, have a way of clashing their wings the one against the other over their backs with a loud snap; another variety called tumblers turn themselves over in the air. Some birds have movements peculiar to the season of love: thus ring-doves, though strong and rapid at other times, yet in the spring hang about on the wing in a toying and playful manner; thus the cock-snipe, while breeding, forgetting his former flight, fans the air like the wind-hover; and the green-finch in particular exhibits such languishing and faltering gestures as to appear like a wounded and dying bird; the king-fisher darts along like an arrow; fern-owls, or goat-suckers, glance in the dusk over the tops of trees like a meteor; starlings as it were swim along, while missel-thrushes use a wild and desultory flight; swallows sweep over the surface of the ground and water, and distinguish themselves by rapid turns and quick evolutions; swifts dash round in circles; and the bank-martin moves with frequent vacillations like a butterfly. Most of the small birds fly by jerks, rising and falling as they advance. Most small birds hop; but wagtails and larks walk, moving their legs alternately. Skylarks rise and fall perpendicularly as they sing: woodlarks hang poised in the air; and titlarks rise and fall in large curves, singing in their descent. The white-throat uses odd jerks and gesticulations over the tops of hedges and bushes. All the duck-kind waddle; divers and auks walk as if fettered, and stand erect on their tails: these are the *compedes* of Linnæus. Geese and cranes, and most wild-fowls, move in figured flights, often changing their position. The secondary remiges of *tringæ*,[161] wild-ducks, and some others, are very long, and give their wings, when the motion, an hooked appearance. Dab-chicks, moor-hens, and coots, fly erect, with their legs hanging down, and hardly make any dispatch; the reason is plain, their wings are placed too forward out of the true centre of gravity; as the legs of auks and divers are situated too backward.

LETTER XLIII

Selborne, Sept. 9, 1778.

DEAR SIR,

From the motion of birds, the transition is natural enough to their notes and language, of which I shall say something. Not that I would pretend to understand their language like the vizer who, by the recital of a conversation which passed between two owls, reclaimed a sultan,* before delighting in conquest and devastation; but I would be thought only to mean that many of the winged tribes have various sounds and voices adapted to express their various passions, wants, and feelings; such as anger, fear, love, hatred, hunger and the like. All species are not equally eloquent; some are copious and fluent as it were in their utterance, while others are confined to a few important sounds: no bird, like the fish kind, is quite mute, though some are rather silent. The language of birds is very ancient, and, like other ancient modes of speech, very elliptical: little is said, but much is meant and understood.

The notes of the eagle-kind are shrill and piercing; and about the season of nidification much diversified, as I have been often assured by a curious observer of nature, who long resided at Gibraltar, where eagles abound. The notes of our hawks much resemble those of the king of birds. Owls have very expressive notes; they hoot in a fine vocal sound, much resembling the *vox humana*, and reducible by a pitch-pipe to a musical key. This note seems to express complacency and rivalry among the males; they use also a quick call and an horrible scream; and can snore and hiss when they mean to menace. Ravens, beside their loud croak, can exert a deep and solemn note that makes the woods to echo; the amorous sound of a crow is strange and ridiculous; rooks, in the breeding season, attempt sometimes

* See *Spectator*, Vol. VII., No. 512.

in the gaiety of their hearts to sing, but with no great success; the parrot-kind have many modulations of voice, as appears by their aptitude to learn human sounds; doves coo in an amorous and mournful manner, and are emblems of despairing lovers; the wood-pecker sets up a sort of loud and hearty laugh; the fern-owl, or goat-sucker, from the dusk till day-break, serenades his mate with the clattering of castanets. All the tuneful *passeres* express their complacency by sweet modulations, and a variety of melody. The swallow, as has been observed in a former letter, by a shrill alarm bespeaks the attention of the other *hirundines*, and bids them be aware that the hawk is at hand. Aquatic and gregarious birds, especially the nocturnal, that shift their quarters in the dark, are very noisy and loquacious; as cranes, wild-geese, wild-ducks, and the like; their perpetual clamour prevents them from dispersing and losing their companions.

In so extensive a subject, sketches and outlines are as much as can be expected; for it would be endless to instance in all the infinite variety of the feathered nation. We shall therefore confine the remainder of this letter to the few domestic fowls of our yards, which are most known, and therefore best understood. At first the peacock, with his gorgeous train, demands our attention; but, like most of the gaudy birds, his notes are grating and shocking to the ear: the yelling of cats, and the braying of an ass, are not more disgustful. The voice of the goose is trumpet-like, and clanking; and once saved the Capitol at Rome, as grave historians assert: the hiss also of the gander is formidable and full of menace, and 'protective of his young.' Among ducks the sexual distinction of voice is remarkable; for, while the quack of the female is loud and sonorous, the voice of the drake is inward and harsh and feeble, and scarce discernible. The cock turkey struts and gobbles to his mistress in a most uncouth manner; he hath also a pert and petulant note when he attacks his adversary. When a hen turkey leads forth her young brood she keeps a watchful eye: and if a bird of prey appear, though ever so high in the air, the careful mother announces the enemy with a little inward moan, and watches him with a steady and attentive look; but if he approach, her note becomes earnest and alarming, and her outcries are redoubled.

No inhabitants of a yard seem possessed of such a variety of expression and so copious a language as common poultry. Take a chicken of four or five days old, and hold it up to a window where there are flies, and it will immediately seize its prey, with little twitterings of complacency; but if you tender it a wasp or a bee, at once its note becomes harsh, and expressive of disapprobation and a sense of danger. When a pullet is ready to lay she intimates the event by a joyous and easy soft note. Of all the occurrences of their life that of laying seems to be the most important; for no sooner has a hen disburdened herself, than she rushes forth with a clamorous kind of joy, which the cock and the rest of his mistresses immediately adopt. The tumult is not confined to the family concerned, but catches from yard to yard, and spreads to every homestead within hearing, till at last the whole village is in an uproar. As soon as a hen becomes a mother her new relation demands a new language; she then runs clocking and screaming about, and seems agitated as if possessed. The father of the flock has also a considerable vocabulary; if he finds food, he calls a favourite concubine to partake; and if a bird of prey passes over, with a warning voice he bids his family beware. The gallant chanticleer has, at command, his amorous phrases, and his terms of defiance. But the sound by which he is best known is his crowing: by this he has been distinguished in all ages as the countryman's clock or larum, as the watchman that proclaims the divisions of the night. Thus the poet elegantly styles him:

> . . . the crested cock, whose clarion sounds
> The silent hours.

A neighbouring gentleman one summer had lost most of

his chickens by a sparrow-hawk, that came gliding down between a faggot-pile and the end of his house to the place where the coops stood. The owner, inwardly vexed to see his flock thus diminishing, hung a setting net adroitly between the pile and the house, into which the caitiff dashed and was entangled. Resentment suggested the law of retaliation: he therefore clipped the hawk's wings, cut off his talons, and, fixing a cork on his bill, threw him down among the brood-hens. Imagination cannot paint the scene that ensued; the expressions that fear, rage, and revenge inspired, were new, or at least such as had been unnoticed before: the exasperated matrons upbraided, they execrated, they insulted, they triumphed. In a word, they never desisted from buffeting their adversary till they had torn him in an hundred pieces.

LETTER XLIV[162]

Selborne.

... monstrent.

*　　*　　*　　*　　*

Quid tantùm Oceano properent se tingere soles
Hyberni; vel quæ tardis mora noctibus obstet.

GENTLEMEN who have outlets might contrive to make ornament subservient to utility; a pleasing eye-trap might also contribute to promote science: an obelisk in a garden or park might be both an embellishment and an heliotrope.

Any person that is curious, and enjoys the advantage of a good horizon, might, with little trouble, make two heliotropes; the one for the winter, the other for the summer solstice: and these two erections might be constructed with very little expense; for two pieces of timber frame-work, about ten or twelve feet high, and four feet broad at the base, and close lined with plank, would answer the purpose.

The erection for the former should, if possible, be placed within sight of some window in the common sitting parlour; because men, at that dead season of the year, are usually within doors at the close of the day; while that for the latter might be fixed for any given spot in the garden or outlet: whence the owner might contemplate, in a fine summer's evening, the utmost extent that the sun makes to the northward at the season of the longest days. Now nothing would be necessary but to place these two objects with so much exactness, that the westerly limb of the sun, at setting, might but just clear the winter heliotrope to the west of it on the shortest day; and that the whole disc of the sun, at the longest day, might exactly at setting also clear the summer heliotrope to the north of it.

By this simple expedient it would soon appear that there is no such thing, strictly speaking, as a solstice; for, from the shortest day, the owner would, every clear evening, see the disc advancing, at its setting, to the westward of the object; and, from the longest day, observe the sun retiring backwards every evening at its setting, towards the object westward, till, in a few nights, it would set quite behind it, and so by degrees to the west of it: for when the sun comes near the summer solstice, the whole disc of it would at first set behind the object: after a time the northern limb would first appear, and so every night gradually more, till at length the whole diameter would set north of it for about three nights; but on the middle night of the three, sensibly more remote than the former or following. When beginning its recess from the summer tropic, it would continue more and more to be hidden every night, till at length it would descend quite behind the object again; and so nightly more and more to the westward.

LETTER XLV

Selborne.

... Mugire videbis
Sub pedibus terram, et descendere montibus ornos.

WHEN I was a boy I used to read, with astonishment and implicit assent, accounts in *Baker's Chronicle*[163] of walking hills and travelling mountains. John Philips, in his *Cyder*,[164] alludes to the credit that was given to such stories with a delicate but quaint vein of humour peculiar to the author of the *Splendid Shilling.*

> I nor advise, nor reprehend the choice
> Of Marcley Hill: the apple no where finds
> A kinder mould: yet 'tis unsafe to trust
> Deceitful ground: who knows but that once more
> This mount may journey, and his present site
> Forsaken, to thy neighbour's bounds transfer
> Thy goodly plants, affording matter strange
> For law debates!

But, when I came to consider better, I began to suspect that though our hills may never have journeyed that far, yet the ends of many of them have slipped and fallen away at distant periods, leaving the cliffs bare and abrupt. This seems to have been the case with Nore and Whetham hills; and especially with the ridge between Harteley Park and Ward-le-ham, where the ground has slid into vast swellings and furrows; and lies still in such romantic confusion as cannot be accounted for from any other cause. A strange event that happened not long since, justifies our suspicions; which, though it befell not within the limits of this parish, yet as it was within the hundred of Selborne, and as the circumstances were singular, may fairly claim a place in a work of this nature.

The months of January and February, in the year 1774, were

remarkable for great melting snows and vast gluts of rain, so that by the end of the latter month the land-springs, or lavants, began to prevail, and to be near as high as in the memorable winter of 1764. The beginning of March also went on in the same tenor; when, in the night between the 8th and 9th of that month, a considerable part of the great woody hanger at Hawkley was torn from its place, and fell down, leaving a high freestone cliff naked and bare, and resembling the steep side of a chalk-pit. It appears that this huge fragment, being perhaps sapped and undermined by waters, foundered, and was engulfed, going down in a perpendicular direction; for a gate which stood in the field, on the top of the hill, after sinking with its posts for thirty or forty feet, remained in so true and upright a position as to open and shut with great exactness, just as in its first situation. Several oaks also are still standing, and in a state of vegetation, after taking the same desperate leap. That great part of this prodigious mass was absorbed in some gulf below, is plain also from the inclining ground at the bottom of the hill, which is free and unincumbered; but would have been buried in heaps of rubbish, had the fragment parted and fallen forward. About an hundred yards from the foot of this hanging coppice stood a cottage by the side of a lane; and two hundred yards lower, on the other side of the lane, was a farm-house, in which lived a labourer and his family; and, just by, a stout new barn. The cottage was inhabited by an old woman and her son and his wife. These people in the evening, which was very dark and tempestuous, observed that the brick floors of their kitchens began to heave and part; and that the walls seemed to open, and the roofs to crack: but they all agree that no tremor of the ground, indicating an earthquake, was ever felt; only that the wind continued to make a most tremendous roaring in the woods and hangers. The miserable inhabitants, not daring to go to bed, remained in the utmost solicitude and confusion, expecting every moment to be buried under the ruins of their shattered edifices. When day-light came they were at leisure to contemplate the devastations of the night: they then found that a deep rift, or chasm, had opened under their houses, and torn them, as it were, in two; and that one end of the barn had suffered in a similar manner; that a pond near the cottage had undergone a strange reverse, becoming deep at the shallow end, and so *vice versa*; that many large oaks were removed out of their perpendicular, some thrown down, and some fallen into the heads of neighbouring trees; and that a gate was thrust forward, with its hedge, full six feet, so as to require a new track to be made to it. From the foot of the cliff the general course of the ground, which is pasture, inclines in a moderate descent for half a mile, and is interspersed with some hillocks, which were rifted, in every direction, as well towards the great woody hanger, as from it. In the first pasture the deep clefts began: and running across the lane, and under the buildings, made such vast shelves that the road was impassable for some time; and so over to an arable field on the other side, which was strangely torn and disordered. The second pasture field, being more soft and springy, was protruded forward without many fissures in the turf, which was raised in long ridges resembling graves, lying at right angles to the motion. At the bottom of this enclosure the soil and turf rose many feet against the bodies of some oaks that obstructed their farther course and terminated this awful commotion.

The perpendicular height of the precipice, in general, is twenty-three yards; the length of the lapse, or slip, as seen from the fields below, one hundred and eighty-one; and a partial fall, concealed in the coppice, extends seventy yards more: so that the total length of this fragment that fell was two hundred and fifty-one yards. About fifty acres of land suffered from this violent convulsion; two houses were entirely destroyed; one end of a new barn was left in ruins, the walls being cracked through the very stones that composed them; a hanging coppice was changed to a naked rock; and some grass grounds and an arable field so broken and rifted by the chasms as to be rendered, for a time, neither fit for the plough or safe for pasturage, till considerable labour and expense had been bestowed in levelling the surface and filling in the gaping fissures.

LETTER XLVI

Selborne.

... resonant arbusta ...

THERE is a steep abrupt pasture field interspersed with furze close to the back of this village, well known by the name of the Short Lithe, consisting of a rocky dry soil, and inclining to the afternoon sun. This spot abounds with the *gryllus campestris*, or field-cricket; which, though frequent in these parts, is by no means a common insect in many other countries.

As their cheerful summer cry cannot but draw the attention of a naturalist, I have often gone down to examine the œconomy of these *grylli*, and study their mode of life: but they are so shy and cautious that it is no easy matter to get a sight of them; for, feeling a person's footsteps as he advances, they stop short in the midst of their song, and retire backward nimbly into their burrows, where they lurk till all suspicion of danger is over.

At first we attempted to dig them out with a spade, but without any great success; for either we could not get to the bottom of the hole, which often terminated under a great stone; or else, in breaking up the ground, we inadvertently squeezed the poor insect to death. Out of one so bruised we took a multitude of eggs, which were long and narrow, of a yellow colour, and covered with a very tough skin. By this accident we learned to distinguish the male from the female; the former of which is shining black, with a golden stripe across his shoulders; the latter is more dusky, more capacious about the abdomen, and carries a long sword-shaped weapon at her tail, which probably is the instrument with which she deposits her eggs in crannies and safe receptacles.

Where violent methods will not avail, more gentle means will often succeed; and so it proved in the present case; for, though a spade be too boisterous and rough an implement, a pliant stalk of grass, gently insinuated into the caverns, will probe their windings to the bottom, and quickly bring out the inhabitant; and thus the humane inquirer may gratify his curiosity without injuring the object of it. It is remarkable that, though these insects are furnished with long legs behind, and brawny thighs for leaping, like grasshoppers; yet when driven from their holes they show no activity, but crawl along in a shiftless manner, so as easily to be taken; and again, though provided with a curious apparatus of wings, yet they never exert them when there seems to be the greatest occasion. The males only make that shrilling noise perhaps out of rivalry and emulation, as is the case with many animals which exert some sprightly note during their breeding time: it is raised by a brisk friction of one wing against the other. They are solitary beings, living singly male or female, each as it may happen: but there must be a time when the sexes have some intercourse, and then the wings may be useful perhaps during the hours of night. When the males meet they will fight fiercely, as I found by some which I put into the crevices of a dry stone wall, where I should have been glad to have made them settle. For though they seemed distressed by being taken out of their knowledge, yet the first that got possession of the chinks would seize upon any that were obtruded upon them with a vast row of serrated fangs. With their strong jaws, toothed like the shears of a lobster's claws, they perforate and round their curious regular cells, having no fore-claws to dig, like the mole-cricket. When taken in hand I could not but wonder that they never offered to defend themselves, though armed with such formidable weapons. Of such herbs as grow before the mouths of their burrows they eat indiscriminately; and on a little platform, which they make just by, they drop their dung; and never, in the day-time, seem to stir more than two or three inches from home. Sitting in the entrance of their caverns they chirp all night as well as day from the middle of the month of May to the middle of July; and in hot weather, when they are most vigorous, they make the hills echo; and, in the stiller hours of darkness, may be heard to a considerable distance. In the beginning of the season, their notes are more faint and inward; but become louder as the summer advances, and so die away again by degrees.

Sounds do not always give us pleasure according to their sweetness and melody; nor do harsh sounds always displease. We are more apt to be captivated or disgusted with the associations which they promote, than with the notes themselves. Thus the shrilling of the field-cricket, though sharp and stridulous, yet marvellously delights some hearers, filling their minds with a train of summer ideas of everything that is rural, verdurous, and joyous.

About the tenth of March the crickets appear at the mouths of their cells, which they then open and bore, and shape very elegantly. All that ever I have seen at that season were in their pupa state, and had only the rudiments of wings, lying under a skin or coat, which must be cast before the insect can arrive at its perfect state;* from whence I should suppose that the old ones of last year do not always survive the winter. In August their holes begin to be obliterated, and the insects are seen no more till spring.

Not many summers ago I endeavoured to transplant a colony to the terrace in my garden, by boring deep holes in the sloping turf. The new inhabitants stayed some time, and fed and sung; but wandered away by degrees, and were heard at a farther distance every morning; so that it appears that on this

---

* We have observed that they cast these skins in April, which are then seen lying at the mouths of their holes.

emergency they made use of their wings in attempting to return to the spot from which they were taken.

One of these crickets, when confined in a paper cage and set in the sun, and supplied with plants moistened with water, will feed and thrive, and become so merry and loud as to be irksome in the same room where a person is sitting: if the plants are not wetted it will die.

LETTER XLVII

Selborne.

Far from all resort of mirth
Save the cricket on the hearth.
MILTON'S *Il Penseroso.*

DEAR SIR,

While many other insects must be sought after in fields and woods, and waters, the *gryllus domesticus*, or house-cricket, resides altogether within our dwellings, intruding itself upon our notice whether we will or no. This species delights in new-built houses, being, like the spider, pleased with the moisture of the walls; and besides, the softness of the mortar enables them to burrow and mine between the joints of the bricks or stones, and to open communications from one room to another. They are particularly fond of kitchens and bakers' ovens, on account of their perpetual warmth.

Tender insects that live abroad either enjoy only the short period of one summer, or else doze away the cold uncomfortable months in profound slumbers; but these, residing as it were in a torrid zone, are always alert and merry: a good Christmas fire is to them like the heats of the dog-days. Though they are frequently heard by day, yet is their natural time of motion only in the night. As soon as it grows dusk, the chirping increases, and they come running forth, and are from the size of a flea to that of their full stature. As one should suppose, from the burning atmosphere which they inhabit, they are a thirsty race, and show a great propensity for liquids, being found frequently drowned in pans of water, milk, broth, or the like. Whatever is moist they affect; and therefore often gnaw holes in wet woollen stockings and aprons that are hung to the fire: they are the housewife's barometer, foretelling her when it will rain; and are prognostic sometimes, she thinks, of ill or good luck; of the death of a near relation, or the approach of an absent lover. By being the constant companions of her solitary hours they naturally become the objects of her superstition. These crickets are not only very thirsty, but very voracious; for they will eat the scummings of pots, and yeast, salt, and crumbs of bread; and any kitchen offal or sweepings. In the summer we have observed them to fly, when it became dusk, out of the windows, and over the neighbouring roofs. This feat of activity accounts for the sudden manner in which they often leave their haunts, as it docs for the method by which they come to houses where they were not known before. It is remarkable, that many sorts of insects seem never to use their wings but when they have a mind to shift their quarters and settle new colonies. When in the air they move '*volatu undoso*,' in waves or curves, like wood-peckers, opening and shutting their wings at every stroke, and so are always rising or sinking.

When they increase to a great degree, as they did once in the house where I am now writing, they become noisome pests, flying into the candles, and dashing into people's faces; but may be blasted and destroyed by gunpowder discharged into their crevices and crannies. In families, at such times, they are, like Pharaoh's plague of frogs, 'in their bed-chambers, and upon their beds, and in their ovens, and in their kneading-troughs.'* Their shrilling noise is occasioned by a brisk attrition of their wings. Cats catch hearth-crickets, and, playing with them as they do with mice, devour them. Crickets may be

*Exod. viii. 3.

destroyed, like wasps, by phials half-filled with beer, or any liquid, and set in their haunts; for, being always eager to drink, they will crowd in till the bottles are full.

LETTER XLVIII

Selborne.

How diversified are the modes of life not only of incongruous but even of congenerous animals; and yet their specific distinctions are not more various than their propensities. Thus, while the field-cricket delights in sunny dry banks, and the house-cricket rejoices amidst the glowing heat of the kitchen hearth or oven, the *gryllus gryllotalpa* (the mole-cricket) haunts moist meadows, and frequents the sides of ponds and banks of streams, performing all its functions in a swampy wet soil. With a pair of fore-feet, curiously adapted to the purpose, it burrows and works under ground like the mole, raising a ridge as it proceeds, but seldom throwing up hillocks.

As mole-crickets often infest gardens by the sides of canals, they are unwelcome guests to the gardener, raising up ridges in their subterraneous progress, and rendering the walks unsightly. If they take to the kitchen quarters, they occasion great damage among the plants and roots, by destroying whole beds of cabbages, young legumes, and flowers. When dug out they seem very slow and helpless, and make no use of their wings by day; but at night they come abroad, and make long excursions, as I have been convinced by finding stragglers, in a morning, in improbable places. In fine weather, about the middle of April, and just at the close of day, they begin to solace themselves with a low, dull, jarring note, continued for a long time without interruption, and not unlike the chattering of the fern-owl, or goat-sucker, but more inward.

About the beginning of May they lay their eggs, as I was once an eye-witness: for a gardener at an house, where I was on a visit, happening to be mowing, on the 6th of that month, by the side of a canal, his scythe struck too deep, pared off a large piece of turf, and laid open to view a curious scene of domestic œconomy:

> ... ingentem lato dedit ore fenestram:
> Apparet domus intus, et atria longa patescunt:
> Apparent ... penetralia.

There were many caverns and winding passages leading to a kind of chamber, neatly smoothed and rounded, and about the size of a moderate snuff-box. Within this secret nursery were deposited near an hundred eggs of a dirty yellow colour, and enveloped in a tough skin, but too lately excluded to contain any rudiments of young, being full of a viscous substance. The eggs lay but shallow, and within the influence of the sun, just under a little heap of fresh-moved mould, like that which is raised by ants.

When mole-crickets fly they move 'cursu undoso,' rising and falling in curves, like the other species mentioned before. In different parts of this kingdom people call them fen-crickets, churr-worms, and eve-churrs, all very apposite names.

Anatomists, who have examined the intestines of these insects, astonish me with their accounts; for they say that, from the structure, position, and number of their stomachs, or maws, there seems to be good reason to suppose that this and the two former species ruminate or chew the cud like many quadrupeds![165]

LETTER XLIX

Selborne, May 7, 1779.

IT is now more than forty years that I have paid some attention to the ornithology of this district, without being able to exhaust the subject: new occurrences still arise as long as any inquiries are kept alive.

In the last week of last month five of those most rare birds, too uncommon to have obtained an English name, but known to naturalists by the terms of *himantopus*, or *loripes*, and *charadrius himantopus*,[166] were shot upon the verge of Frinsham-pond, a large lake belonging to the bishop of Winchester, and lying between Wolmer-forest, and the town of Farnham, in the county of Surrey. The pond keeper says there were three brace in the flock; but that, after he had satisfied his curiosity, he suffered the sixth to remain unmolested. One of these specimens I procured, and found the length of the legs to be so extraordinary, that, at first sight, one might have supposed the shanks had been fastened on to impose on the credulity of the beholder: they were legs in *caricatura*; and had we seen such proportions on a Chinese or Japan screen we should have made large allowances for the fancy of the draughtsman. These birds are of the plover family, and might with propriety be called the stilt plovers. Brisson,[167] under that idea, gives them the apposite name of *l'échasse*. My specimen, when drawn and stuffed with pepper, weighed only four ounces and a quarter, though the naked part of the thigh measured three inches and an half, and the legs four inches and an half. Hence we may safely assert that these birds exhibit, weight for inches, incomparably the greatest length of legs of any known bird. The flamingo, for instance, is one of the most long legged birds, and yet it bears no manner of proportion to the *himantopus;* for a cock flamingo weighs, at an average, about four pounds avoirdupois; and his legs and thighs measure usually about twenty inches. But four pounds are fifteen times and a fraction more than four ounces and one quarter; and if four ounces and a quarter have eight inches of legs, four pounds must have one hundred and twenty inches and a fraction of legs; viz., somewhat more than ten feet; such a monstrous proportion as the world never saw! If you should try the experiment in still larger birds the disparity would still increase. It must be matter of great curiosity to see the stilt plover move; to observe how it can wield such a length of lever with such feeble muscles as the thighs seem to be furnished with. At best one should expect it to be but a bad walker: but what adds to the wonder is that it has no back toe. Now without that steady prop to support its steps it must be liable, in speculation, to perpetual vacillations, and seldom able to preserve the true centre of gravity.

The old name of *himantopus* is taken from Pliny; and, by an awkward metaphor, implies that the legs are as slender and pliant as if cut out of a thong of leather. Neither Willughby nor Ray, in all their curious researches either at home or abroad, ever saw this bird. Mr Pennant never met with it in all Great Britain, but observed it often in the cabinets of the curious at Paris. Hasselquist says that it migrates to Egypt in the autumn: and a most accurate observer of nature has assured me that he has found it on the banks of the streams in Andalusia.

Our writers record it to have been found only twice in Great Britain. From all these relations it plainly appears that these long-legged plovers are birds of South Europe, and rarely visit our island; and when they do are wanderers and stragglers, and impelled to make so distant and northern an excursion from motives or accidents for which we are not able to account. One thing may fairly be deduced, that these birds come over to us from the continent, since nobody can suppose that a species not noticed once in an age, and of such a remarkable make, can constantly breed unobserved in this kingdom.

LETTER L

Selborne, April 21, 1780.

DEAR SIR,

The old Sussex tortoise, that I have mentioned to you so often, is become my property. I dug it out of its winter dormitory in March last, when it was enough awakened to express its resentments by hissing; and, packing it in a box with earth, carried it eighty miles in post-chaises. The rattle and hurry of the journey so perfectly roused it that, when I turned it out on a border, it walked twice down to the bottom of my garden; however, in the evening, the weather being cold, it buried itself in the loose mound, and continues still concealed.

As it will be under my eye, I shall now have an opportunity of enlarging my observations on its mode of life, and propensities; and perceive already that, towards the time of coming forth, it opens a breathing place in the ground near its head, requiring, I conclude, a freer respiration, as it becomes more alive. This creature not only goes under the earth from the middle of November to the middle of April, but sleeps great part of the summer; for it goes to bed in the longest days at four in the afternoon, and often does not stir in the morning till late. Besides, it retires to rest for every shower; and does not move at all in wet days.

When one reflects on the state of this strange being, it is a matter of wonder to find that Providence should bestow such a profusion of days, such a seeming waste of longevity, on a reptile that appears to relish it so little as to squander more than two-thirds of its existence in joyless stupor, and be lost to all sensation for months together in the profoundest of slumbers.

While I was writing this letter, a moist and warm afternoon, with the thermometer at 50, brought forth troups of shell-snails; and, at the same juncture, the tortoise heaved up the mould and put out its head; and the next morning came forth, as it were raised from the dead; and walked about till four in the afternoon. This was a curious coincidence! a very amusing occurrence! to see such a similarity of feelings between the two φερέοικοι! for so the Greeks call both the shell-snail and the tortoise.

Summer birds are, this cold and backward spring, unusually late: I have seen but one swallow yet. This conformity with the weather convinces me more and more that they sleep in the winter.

LETTER LI

Selborne, Sept. 3, 1781.

I HAVE now read your miscellanies through with much care and satisfaction: and am to return you my best thanks for the honourable mention made in them of me as a naturalist, which I wish I may deserve.

In some former letters I expressed my suspicions that many of the house-martins do not depart in the winter far from this village. I therefore determined to make some search about the south-east end of the hill, where I imagined they might slumber out the uncomfortable months of winter. But supposing that the examination would be made to the best advantage in the spring, and observing that no martins had appeared by the 11th of April last; on that day I employed some men to explore the shrubs and cavities of the suspected spot. The persons took pains, but without any success: however, a remarkable incident occurred in the midst of our pursuit – while the labourers were at work a house-martin, the first that had been seen this year, came down the village in the sight of several people, and went at once into a nest, where it stayed a short time, and then flew over the houses; for some days after no martins were observed, not till the 16th of April, and then only a pair. Martins in general were remarkably late this year.

LETTER LII

Selborne, Sept. 9, 1781.

I HAVE just met with a circumstance respecting swifts, which furnishes an exception to the whole tenor of my observations ever since I have bestowed any attention on that species of *hirundines.* Our swifts, in general, withdrew this year about the first day of August, all save one pair, which in two or three days was reduced to a single bird. The perseverance of this individual made me suspect that the strongest of motives, that of an attachment to her young, could alone occasion so late a stay. I watched therefore till the twenty-fourth of August, and then discovered that, under the eaves of the church, she attended upon two young, which were fledged, and now put out their white chins from a crevice. These remained till the twenty-seventh, looking more alert every day, and seeming to long to be on the wing. After this day they were missing at once; nor could I ever observe them with their dam coursing round the church in the act of learning to fly, as the first broods evidently do. On the thirty-first I caused the eaves to be searched, but we found in the nest only two callow, dead, stinking swifts, on which a second nest had been formed. This double nest was full of the black shining cases of the *hippoboscæ hirundinis.*

The following remarks on this unusual incident are obvious. The first is, that though it may be disagreeable to swifts to remain beyond the beginning of August, yet that they can subsist longer is undeniable. The second is, that this uncommon event, as it was owing to the loss of the first brood, so it corroborates my former remark, that swifts breed regularly but once; since, was the contrary the case, the occurrence above could neither be new nor rare.

P.S. One swift was seen at Lyndon, in the county of Rutland, in 1782, so late as the third of September.

LETTER LIII

As I have sometimes known you make inquiries about several kinds of insects, I shall here send you an account of one sort which I little expected to have found in this kingdom. I had often observed that one particular part of a vine growing on the walls of my house was covered in the autumn with a black dust-like appearance, on which the flies fed eagerly; and that the shoots and leaves thus affected did not thrive; nor did the fruit ripen. To this substance I applied my glasses; but could not discover that it had anything to do with animal life, as I at first expected: but, upon a closer examination behind the larger boughs, we were surprised to find that they were coated over with husky shells, from whose sides proceeded a cotton-like substance, surrounding a multitude of eggs. This curious and uncommon production put me upon recollecting what I have heard and read concerning the *coccus vitis viniferæ*[168] of Linnæus, which, in the South of Europe, infests many vines, and is an horrid and loathsome pest. As soon as I had turned to the accounts given of this insect, I saw at once that it swarmed on my vine; and did not appear to be at all checked by the preceding winter, which had been uncommonly severe.

Not being then at all aware that it had anything to do with England, I was much inclined to think that it came from Gibraltar among the many boxes and packages of plants and birds which I had formerly received from thence; and especially as the vine infested grew immediately under my study-window, where I usually kept my specimens. True it is that I had received nothing from thence for some years: but as insects, we know, are conveyed from one country to another in a very unexpected manner, and have a wonderful power of maintaining their existence till they fall into a *nidus* proper for their support and increase, I cannot but suspect still that these *cocci* came to me originally from Andalusia. Yet, all the while, candour obliges me to confess that Mr Lightfoot[169] has written me word that he once, and but once, saw these insects on a vine at Weymouth in Dorsetshire; which, it is here to be observed, is a seaport town to which the *coccus* might be conveyed by shipping.

As many of my readers may possibly never have heard of this strange and unusual insect, I shall here transcribe a passage from a natural history of Gibraltar, written by the Reverend John White, late vicar of Blackburn in Lancashire, but not yet published:

'In the year 1770 a vine which grew on the east side of my house, and which had produced the finest crops of grapes for years past, was suddenly overspread on all the woody branches with large lumps of a white fibrous substance resembling spiders' webs, or rather raw cotton. It was of a very clammy quality, sticking fast to everything that touched it, and capable of being spun into long threads. At first I suspected it to be the product of spiders, but could find none. Nothing was to be seen connected with it but many brown oval husky shells, which by no means looked like insects, but rather resembled bits of the dry bark of the vine. The tree had a plentiful crop of grapes set, when this pest appeared upon it; but the fruit was manifestly injured by this foul incumbrance. It remained all the summer, still increasing, and loaded the woody and bearing branches to a vast degree. I often pulled off great quantities by handfuls; but it was so slimy and tenacious that it could by no means be cleared. The grapes never filled to their natural perfection, but turned watery and vapid. Upon perusing the works afterwards of M. de Réaumur, I found this matter perfectly described and accounted for. Those husky shells, which I had observed, were no other than the female *coccus*, from whose sides this cotton-like substance exudes, and serves as a covering and security for their eggs.'

To this account I think proper to add, that, though the female *cocci* are stationary, and seldom remove from the place to which they stick, yet the male is a winged insect; and that the black dust which I saw was undoubtedly the excrement of the females, which is eaten by ants as well as flies. Though the utmost severity of our winter did not destroy these insects, yet the attention of the gardener in a summer or two has entirely relieved my vine from this filthy annoyance.

As we have remarked above that insects are often conveyed from one country to another in a very unaccountable manner, I shall here mention an emigration of small aphides, which was observed in the village of Selborne no longer ago than August the 1st, 1785.

At about three o'clock in the afternoon of that day, which was very hot, the people of this village were surprised by a shower of aphides, or smother-flies, which fell in these parts. Those that were walking in the street at that juncture found themselves covered with these insects, which settled also on the hedges and gardens, blackening all the vegetables where they alighted. My annuals were discoloured with them, and the stalks of a bed of onions were quite coated over for six days after. These armies were then no doubt, in a state of emigration, and shifting their quarters; and might have come, as far as we know, from the great hop-plantations of Kent or Sussex, the wind being all that day in the easterly quarter. They were observed at the same time in great clouds about Farnham, and all along the vale from Farnham to Alton.*

* For various methods by which several insects shift their quarters, see Derham's *Physico-Theology*.

LETTER LIV

Dear Sir,

When I happen to visit a family where gold and silver fishes are kept in a glass bowl, I am always pleased with the occurrence, because it offers me an opportunity of observing the actions and propensities of those beings with whom we can be little acquainted in their natural state. Not long since I spent a fortnight at the house of a friend where there was such a vivary, to which I paid no small attention, taking every occasion to remark what passed within its narrow limits. It was here that I first observed the manner in which fishes die. As soon as the creature sickens, the head sinks lower and lower, and it stands as it were on its head; till, getting weaker, and losing all poise, the tail turns over, and at last it floats on the surface of the water with its belly uppermost. The reason why fishes, when dead, swim in that manner is very obvious; because, when the body is no longer balanced by the fins of the belly, the broad muscular back preponderates by its own gravity, and turns the belly uppermost, as lighter from its being a cavity, and because it contains the swimming-bladders, which contribute to render it buoyant. Some that delight in gold and silver fishes have adopted a notion that they need no aliment. True it is that they will subsist for a long time without any apparent food but what they can collect from pure water frequently changed; yet they must draw some support from animalcula, and other nourishment supplied by the water; because, though they seem to eat nothing, yet the consequences of eating often drop from them. That they are best pleased with such jejune diet may easily be confuted, since if you toss them crumbs, they will seize them with great readiness, not to say greediness: however, bread should be given sparingly, lest, turning sour, it corrupt the water. They will also feed on the water plant called *lemna* (duck's meat), and also on small fry.

When they want to move a little they gently protrude themselves with their *pinnæ pectorales*; but it is with their strong muscular tails only that they and all fishes shoot along with such inconceivable rapidity. It has been said that the eyes of fishes are immoveable: but these apparently turn them forward or backward in their sockets as their occasions require. They take little notice of a lighted candle, though applied close to their heads, but flounce and seem much frightened by a sudden stroke of the hand against the support whereon the bowl is hung; especially when they have been motionless, and are perhaps asleep. As fishes have no eyelids, it is not easy to discern when they are sleeping or not, because their eyes are always open.

Nothing can be more amusing than a glass bowl containing such fishes: the double refractions of the glass and water represent them, when moving, in a shifting and changeable variety of dimensions, shades, and colours; while the two mediums, assisted by the concavo-convex shape of the vessel, magnify and distort them vastly; not to mention that the introduction of another element and its inhabitants into our parlours engages the fancy in a very agreeable manner.

Gold and silver fishes, though originally natives of China and Japan, yet are become so well reconciled to our climate as to thrive and multiply very fast in our ponds and stews.[170] Linnæus ranks this species of fish under the genus of *cyprinus*, or carp, and calls it *cyprinus auratus*.

Some people exhibit this sort of fish in a very fanciful way; for they cause a glass bowl to be blown with a large hollow space within, that does not communicate with it. In this cavity they put a bird occasionally; so that you may see a goldfinch or a linnet hopping as it were in the midst of the water, and the fishes swimming in a circle round it. The simple exhibition of the fishes is agreeable and pleasant; but in so complicated a way becomes whimsical and unnatural, and liable to the objection due to him,

Qui variare cupit rem prodigialiter unam.

I am, etc.

LETTER LV

October 10, 1781.

DEAR SIR,

I think I have observed before that much the most consider-able part of the house-martins withdraw from hence about the first week in October; but that some, the latter broods I am now convinced, linger on till towards the middle of that month: and that at times, once perhaps in two or three years, a flight, for one day only, has shown itself in the first week of November.

Having taken notice, in October 1780, that the last flight was numerous, amounting perhaps to one hundred and fifty; and that the season was soft and still; I was resolved to pay un-common attention to these late birds; to find, if possible, where they roosted, and to determine the precise time of their retreat. The mode of life of these latter *hirundines* is very favourable to such a design; for they spend the whole day in the sheltered district, between me and the Hanger, sailing about in a placid, easy manner, and feasting on those insects which love to haunt a spot so secure from ruffling winds. As my principal object was to discover the place of their roosting, I took care to wait on them before they retired to rest, and was much pleased to find that, for several evenings together, just at a quarter past five in the afternoon, they all scudded away in great haste towards the south-east, and darted down among the low shrubs above the cottages at the end of the hill. This spot in many respects seems to be well calculated for their winter residence: for in many parts it is as steep as the roof of any house, and therefore secure from the annoyances of water; and it is moreover clothed with beechen shrubs, which, being stunted and bitten by sheep, make the thickest covert imaginable; and are so entangled as to be impervious to the smallest spaniel: besides, it is the nature of underwood beech never to cast its leaf all the winter; so that, with the leaves on the ground and those on the twigs, no shelter can be more complete. I watched them on to the thirteenth and fourteenth of October, and found their evening retreat was exact and uniform; but after this they made no regular appearance. Now and then a straggler was seen; and on the twenty-second of October, I observed two in the morning over the village, and with them my remarks for the season ended.

From all these circumstances put together, it is more than probable that this lingering flight, at so late a season of the year, never departed from the island. Had they indulged me that autumn with a November visit, as I much desired, I presume that, with proper assistants, I should have settled the matter past all doubt; but though the third of November was a sweet day, and in appearance exactly suited to my wishes, yet not a martin was to be seen; and so I was forced, reluctantly, to give up the pursuit.

I have only to add that were the bushes, which cover some acres, and are not my own property, to be grubbed and carefully examined, probably those late broods, and perhaps the whole aggregate body of the house-martins of this district, might be found there, in different secret dormitories; and that so far from withdrawing into warmer climes, it would appear that they never depart three hundred yards from the village.

LETTER LVI

THEY who write on natural history cannot too frequently advert to instinct, that wonderful limited faculty, which, in some instances, raises the brute creation as it were above reason, and in others leaves them so far below it. Philosophers have defined instinct to be that secret influence by which every species is impelled naturally to pursue, at all times, the same way or track, without any teaching or example; whereas reason, without instruction, would often vary and do that by many methods which instinct effects by one alone. Now this maxim must be taken in a qualified sense; for there are instances in which instinct does vary and conform to the circumstances of place and convenience.

It has been remarked that every species of bird has a mode of nidification peculiar to itself; so that a schoolboy would at once pronounce on the sort of nest before him. This is the case among fields and woods, and wilds; but, in the villages round London, where mosses and gossamer, and cotton from vegetables, are hardly to be found, the nest of the chaffinch has not that elegant finished appearance, nor is it so beautifully studded with lichens, as in a more rural district: and the wren is obliged to construct its house with straws and dry grasses, which do not give it that rotundity and compactness so remarkable in the edifices of the little architect. Again, the regular nest of the house-martin is hemispheric; but where a rafter, or a joist, or a cornice may happen to stand in the way, the nest is so contrived as to conform to the obstruction, and becomes flat or oval, or compressed.

In the following instances instinct is perfectly uniform and consistent. There are three creatures, the squirrel, the field-mouse, and the bird called the nut-hatch (*sitta Europæa*), which live much on hazel nuts; and yet they open them each in a different way. The first, after rasping off the small end, splits the shell in two with his long fore-teeth, as a man does with his knife; the second nibbles a hole with his teeth, so regular as if drilled with a wimble,[171] and yet so small that one would wonder how the kernel can be extracted through it; while the last picks an irregular ragged hole with its bill: but as this artist has no paws to hold the nut firm while he pierces it, like an adroit workman, he fixes it, as it were in a vice, in some cleft of a tree, or in some crevice; when, standing over it, he perforates the stubborn shell. We have often placed nuts in the chink of a gate-post where nut-hatches have been known to haunt, and have always found that those birds have readily penetrated them. While at work they make a rapping noise that may be heard at a considerable distance.

You that understand both the theory and practical part of music may best inform us why harmony or melody should so strangely affect some men, as it were by recollection, for days after a concert is over. What I mean the following passage will most readily explain:

'Præhabebat porrò vocibus humanis, instrumentisque harmonicis musicam illam avium: non quod aliâ quoque non delectaretur; sed quod ex musicâ humanâ relinqueretur in animo continens quædam, attentionemque et somnum conturbans agitatio; dum ascensus, exscensus, tenores, ac mutationes illæ sonorum et consonantiarum euntque redeuntque per phantasiam: cum nihil tale relinqui possit ex modulationibus avium, quæ, quod non sunt perinde a nobis imitabiles, non possunt perinde internam facultatem commovere.'[172] – GASSENDUS in Vitâ Peireskii.

This curious quotation strikes me much by so well representing my own case, and by describing what I have so often felt, but never could so well express. When I hear fine music I am haunted with passages therefrom night and day; and especially at first waking, which, by their importunity, give me more uneasiness than pleasure: elegant lessons still tease my imagination, and recur irresistibly to my recollection at seasons, and even when I am desirous of thinking of more serious matters.

I am, etc.

LETTER LVII

A RARE, and I think a new little bird frequents my garden, which I have great reason to think is the pettichaps:[173] it is common in some parts of the kingdom; and I have received formerly several dead specimens from Gibraltar. This bird much resembles the white-throat, but has a more white or rather silvery breast and belly; is restless and active, like the willow-wrens, and hops from bough to bough, examining every part for food; it also runs up the stems of the crown-imperials, and, putting its head into the bells of those flowers, sips the liquor which stands in the nectarium of each petal. Sometimes it feeds on the ground, like the hedge-sparrow, by hopping about on the grass-plots and mown walks.

One of my neighbours, an intelligent and observing man, informs me that, in the beginning of May, and about ten minutes before eight o'clock in the evening, he discovered a great cluster of house-swallows, thirty at least he supposes, perching on a willow that hung over the verge of James Knight's upper-pond. His attention was first drawn by the twittering of these birds, which sat motionless in a row on the bough, with their heads all one way, and, by their weight, pressing down the twig so that it nearly touched the water. In this situation he watched them till he could see no longer. Repeated accounts of this sort, spring and fall, induce us greatly to suspect that house-swallows have some strong attachment to water, independent of the matter of food; and though they may not retire into that element, yet they may conceal themselves in the banks of pools and rivers during the uncomfortable months of winter.

One of the keepers of Wolmer-forest sent me a peregrine falcon, which he shot on the verge of that district as it was devouring a wood-pigeon. The *falco peregrinus*, or haggard falcon, is a noble species of hawk seldom seen in the southern counties. In winter 1767[174] one was killed in the neighbouring parish of Faringdon, and sent by me to Mr Pennant into North Wales.* Since that time I have met with none till now. The specimen measured above was in fine preservation, and not injured by the shot: it measured forty-two inches from wing to wing, and twenty-one from beak to tail, and weighed two pounds and an half standing weight. This species is very robust, and wonderfully formed for rapine: its breast was plump and muscular; its thighs long, thick, and brawny; and its legs remarkably short and well set: the feet were armed with most formidable, sharp, long talons: the eyelids and cere of the bill were yellow; but the irides of the eyes dusky; the beak was thick and hooked, and of a dark colour, and had a jagged process near the end of the upper mandible on each side: its tail, or train, was short in proportion to the bulk of its body: yet the wings, when closed, did not extend to the end of the train. From its large and fair proportions it might be supposed to have been a female; but I was not permitted to cut open the specimen. For one of the birds of prey, which are usually lean, this was in high case: in its craw were many barleycorns, which probably came from the crop of the wood-pigeon, on which it was feeding when shot: for voracious birds do not eat grain; but when devouring their quarry, with undistinguishing vehemence swallow bones and feathers, and all matters, indiscriminately. This falcon was probably driven from the mountains of North Wales or Scotland, where they are known to

*See my tenth and eleventh letter to that gentleman.

breed, by rigorous weather and deep snows that had lately fallen.

I am, etc.

LETTER LVIII

My near neighbour, a young gentleman in the service of the East-India Company, has brought home a dog and a bitch of the Chinese breed from Canton; such as are fattened in the country for the purpose of being eaten: they are about the size of a moderate spaniel; of a pale yellow colour, with coarse bristling hairs on their backs; sharp upright ears, and peaked heads, which give them a very fox-like appearance. Their hind legs are unusually straight, without any bend at the hock or ham, to such a degree as to give them an awkward gait when they trot. When they are in motion their tails are curved high over their backs like those of some hounds, and have a bare place each on the outside from the tip midway, that does not seem to be matter of accident, but somewhat singular. Their eyes are jet black, small, and piercing; the insides of their lips and mouths of the same colour, and their tongues blue. The bitch has a dew-claw on each hind leg; the dog has none. When taken out into a field the bitch showed some disposition for hunting, and dwelt on the scent of a covey of partridges till she sprung them, giving her tongue all the time. The dogs in South America are dumb; but these bark much in a short thick manner, like foxes; and have a surly, savage demeanour like their ancestors, which are not domesticated, but bred up in sties, where they are fed for the table with rice-meal and other farinaceous food. These dogs, having been taken on board as soon as weaned, could not learn much from their dam; yet they did not relish flesh when they came to England. In the islands of the Pacific Ocean the dogs are bred up on vegetables, and would not eat flesh when offered them by our circumnavigators.

We believe that all dogs, in a state of nature, have sharp, upright fox-like ears; and that hanging ears, which are esteemed so graceful, are the effect of choice breeding and cultivation. Thus, in the Travels of Ysbrandt Ides from Muscovy to China, the dogs which draw the Tartars on snow-sledges near the river Oby are engraved with prick-ears, like those from Canton. The Kamschatdales also train the same sort of sharp-eared peak-nosed dogs to draw their sledges; as may be seen in an elegant print engraved for Captain Cook's last voyage round the world.

Now we are upon the subject of dogs it may not be impertinent to add, that spaniels, as all sportsmen know, though they hunt partridges and pheasants as it were by instinct, and with much delight and alacrity, yet will hardly touch their bones when offered as food; nor will a mongrel dog of my own, though he is remarkable for finding that sort of game. But, when we came to offer the bones of partridges to the two Chinese dogs, they devoured them with much greediness, and licked the platter clean.

No sporting dogs will flush woodcocks till inured to the scent and trained to the sport, which they then pursue with vehemence and transport; but then they will not touch their bones, but turn from them with abhorrence, even when they are hungry.

Now, that dogs should not be fond of the bones of such birds as they are not disposed to hunt is no wonder; but why they reject and do not care to eat their natural game is not so easily accounted for, since the end of hunting seems to be, that the chase pursued should be eaten. Dogs again will not devour the more rancid water-fowls, nor indeed the bones of any wild-fowls; nor will they touch the fœtid bodies of birds that feed on offal and garbage: and indeed there may be somewhat of providential instinct in this circumstance of dislike; for vultures,* and kites, and ravens, and crows, etc., were intended to be messmates with dogs† over their carrion; and seem to be appointed by nature as fellow-scavengers to remove all cadaverous nuisances from the face of the earth.

I am, etc.

LETTER LIX

THE fossil wood buried in the bogs of Wolmer-forest is not yet all exhausted, for the peat-cutters now and then stumble upon a log. I have just seen a piece which was sent by a labourer of Oakhanger to a carpenter of this village; this was the butt-end of a small oak, about five feet long, and about five inches in diameter. It had apparently been severed from the ground by an axe, was very ponderous, and as black as ebony. Upon asking the carpenter for what purpose he had procured it, he told me that it was to be sent to his brother, a joiner at Farnham, who was to make use of it in cabinet work, by inlaying it along with whiter woods.

Those that are much abroad on evenings after it is dark, in spring and summer, frequently hear a nocturnal bird passing by on the wing, and repeating often a short quick note. This bird I have remarked myself, but never could make out till lately. I am assured now that it is the stone curlew (*charadrius œdicnemus*). Some of them pass over or near my house almost every evening after it is dark, from the uplands of the hill and

---

* Hasselquist, in his Travels to the Levant, observes that the dogs and vultures at Grand Cairo maintain such a friendly intercourse as to bring up their young together in the same place.

† The Chinese word for a dog to an European ear sounds like *quihloh*.

North field, away down towards Dorton; where, among the streams and meadows, they find a greater plenty of food. Birds that fly by night are obliged to be noisy; their notes often repeated become signals or watchwords to keep them together, that they may not stray or lose each the other in the dark.

The evening proceedings and manœuvres of the rooks are curious and amusing in the autumn. Just before dusk they return in long strings from the foraging of the day, and rendezvous by thousands over Selborne-down, where they wheel round in the air, and sport and dive in a playful manner, all the while exerting their voices, and making a loud cawing, which, being blended and softened by the distance that we at the village are below them, becomes a confused noise or chiding; or rather a pleasing murmur, very engaging to the imagination, and not unlike the cry of a pack of hounds in hollow, echoing woods, or the rushing of the wind in tall trees, or the tumbling of the tide upon a pebbly shore. When this ceremony is over, with the last gleam of day, they retire for the night to the deep beechen woods of Tisted and Ropley. We remember a little girl who, as she was going to bed, used to remark on such an occurrence, in the true spirit of physico-theology, that the rooks were saying their prayers; and yet this child was much too young to be aware that the scriptures have said of the Deity – that 'he feedeth the ravens who call upon him.'

I am, etc.

LETTER LX

In reading Dr Huxham's *Observationes de Aëre*,[175] etc., written at Plymouth, I find by those curious and accurate remarks, which contain an account of the weather from the year 1727 to the year 1748, inclusive, that though there is frequent rain in that district of Devonshire, yet the quantity falling is not great; and that some years it has been very small: for in 1731 the rain measured only 17.266 in. and in 1741, 20.354 in.; and again in 1743 only 20.908 in. Places near the sea have frequent scuds, that keep the atmosphere moist, yet do not reach far up into the country; making thus the maritime situations appear wet, when the rain is not considerable. In the wettest years at Plymouth the Doctor measured only once 36 in.; and again once, viz., 1734, 37.114 in.: a quantity of rain that has twice been exceeded at Selborne in the short period of my observations. Dr Huxham remarks, that frequent small rains keep the air moist; while heavy ones render it more dry, by beating down the vapours. He is also of opinion that the dingy, smoky appearance of the sky, in very dry seasons, arises from the want of moisture sufficient to let the light through, and render the atmosphere transparent; because he had observed several bodies more diaphanous when wet than dry; and did never recollect that the air had that look in rainy seasons.

My friend who lives just beyond the top of the down,[176] brought his three swivel guns to try them in my outlet, with their muzzles towards the Hanger, supposing that the report would have had a great effect; but the experiment did not answer his expectation. He then removed them to the Alcove on the Hanger: when the sound, rushing along the Lythe and Combwood, was very grand: but it was at the Hermitage that the echoes and repercussions delighted the hearers; not only filling the Lythe with the roar, as if all the beeches were tearing up by the roots; but, turning to the left, they pervaded the vale above Combwood-ponds; and after a pause seemed to take up the crash again, and to extend round Harteley-hangers, and to die away at last among the coppices and coverts of Ward le ham. It has been remarked before that this district is an Anathoth, a place of responses or echoes, and therefore proper for such experiments: we may further add that the pauses in echoes, when they cease and yet are taken up again, like the pauses in music, surprise the hearers, and have a fine effect on the imagination.

The gentleman above mentioned has just fixed a barometer in his parlour at Newton Valence. The tube was first filled here (at Selborne) twice with care, when the mercury agreed and stood exactly with my own; but being filled again twice at Newton, the mercury stood, on account of the great elevation of that house, three-tenths of an inch lower than the barometers at this village, and so continues to do, be the weight of the atmosphere what it may. The plate of the barometer at Newton is figured as low as 27; because in stormy weather the mercury there will sometimes descend below 28. We have supposed Newton-house to stand two hundred feet higher than this house: but if the rule holds good, which says that mercury in a barometer sinks one-tenth of an inch for every hundred feet elevation, then the Newton barometer, by standing three-tenths lower than that of Selborne, proves that Newton-house must be three hundred feet higher than that in which I am writing, instead of two hundred.

It may not be impertinent to add, that the barometers at Selborne stand three-tenths of an inch lower than the barometers at South Lambeth; whence we may conclude that the former place is about three hundred feet higher than the latter; and with good reason, because the streams that rise with us run into the Thames at Weybridge, and so to London. Of course therefore there must be lower ground all the way from Selborne to South Lambeth; the distance between which, all the windings and indentings of the streams considered, cannot be less than an hundred miles.

I am, etc.

LETTER LXI

SINCE the weather of a district is undoubtedly part of its natural history, I shall make no further apology for the four following letters, which will contain many particulars concerning some of the great frosts and a few respecting some very hot summers, that have distinguished themselves from the rest during the course of my observations.

As the frost in January 1768 was, for the small time it lasted, the most severe that we had then known for many years, and was remarkably injurious to evergreens, some account of its rigour, and reason of its ravages, may be useful, and not unacceptable to persons that delight in planting and ornamenting; and may particularly become a work that professes never to lose sight of utility.

For the last two or three days of the former year there were considerable falls of snow, which lay deep and uniform on the ground without any drifting, wrapping up the more humble vegetation in perfect security. From the first day to the fifth of the new year more snow succeeded; but from that day the air became entirely clear; and the heat of the sun about noon had a considerable influence in sheltered situations.

It was in such an aspect that the snow on the author's evergreens was melted every day, and frozen intensely every night; so that the laurustines, bays, laurels, and arbutuses looked, in three or four days, as if they had been burnt in the fire; while a neighbour's plantation of the same kind, in a high cold situation, where the snow was never melted at all, remained uninjured.

From hence I would infer that it is the repeated melting and freezing of the snow that is so fatal to vegetation, rather than the severity of the cold. Therefore it highly behoves every planter, who wishes to escape the cruel mortification of losing in a few days the labour and hopes of years, to bestir himself on such emergencies; and, if his plantations are small, to avail himself of mats, cloths, pease-haum, straw, reeds, or any such covering, for a short time; or, if his shrubberies are extensive, to see that his people go about with prongs and forks, and carefully dislodge the snow from the boughs, since the naked foliage will shift much better for itself, than where the snow is partly melted and frozen again.

It may perhaps appear at first like a paradox; but doubtless the more tender trees and shrubs should never be planted in hot aspects; not only for the reason assigned above, but also because, thus circumstanced, they are disposed to shoot earlier in the spring, and grow on later in the autumn than they would otherwise do, and so are sufferers by lagging or early frosts. For this reason also plants from Siberia will hardly endure our climate: because, on the very first advances of spring, they shoot away, and so are cut off by the severe nights of March and April.

Dr Fothergill[177] and others have experienced the same inconvenience with respect to the more tender shrubs from North America; which they therefore plant under north walls. There should also perhaps be a wall to the east to defend them from the piercing blasts from that quarter.

This observation might without any impropriety be carried into animal life; for discerning bee-masters now find that their hives should not in the winter be exposed to the hot sun, because such unseasonable warmth awakens the inhabitants too early from their slumbers; and, by putting their juices into motion too soon, subjects them afterwards to inconveniences when rigorous weather returns.

The coincidents attending this short but intense frost were, that the horses fell sick with an epidemic distemper, which injured the winds of many, and killed some; that colds and coughs were general among the human species; that it froze under people's beds for several nights; that meat was so hard frozen that it could not be spitted, and could not be secured but in cellars; that several redwings and thrushes were killed by the frost; and that the large titmouse continued to pull straw lengthwise from the eaves of thatched houses and barns in a most adroit manner, for a purpose that has been explained already.*

On the 3rd of January, Benjamin Martin's thermometer within doors, in a close parlour where there was no fire, fell in the night to 20, and on the 4th to 18, and the 7th to $17\frac{1}{2}$, a degree of cold which the owner never since saw in the same situation; and he regrets much that he was not able at that juncture to attend his instrument abroad. All this time the wind continued north and north-east; and yet on the eighth roost-cocks, which had been silent, began to sound their clarions, and crows to clamour, as prognostic of milder weather; and, moreover, moles began to heave and work, and a manifest thaw took place. From the latter circumstance we may conclude that thaws often originate under ground from warm vapours which arise; else how should subterraneous animals receive such early intimations of their approach? Moreover, we have often observed that cold seems to descend from above; for, when a thermometer hangs abroad in a frosty night, the intervention of a cloud shall immediately raise the mercury ten degrees; and a clear sky shall again compel it to descend to its former gauge.

And here it may be proper to observe, on what has been said above, that though frosts advance to their utmost severity by somewhat of a regular gradation, yet thaws do not usually come on by as regular a declension of cold; but often take place immediately from intense freezing; as men in sickness often mend at once from a paroxysm.

To the great credit of Portugal laurels and American junipers, be it remembered that they remained untouched amidst the general havoc: hence men should learn to ornament chiefly

*See Letter XLI to Mr Pennant.

with such trees as are able to withstand accidental severities, and not subject themselves to the vexation of a loss which may befall them once perhaps in ten years, yet may hardly be recovered through the whole course of their lives.

As it appeared afterwards the ilexes were much injured, the cypresses were half destroyed, the arbutuses lingered on, but never recovered; and the bays, laurustines, and laurels, were killed to the ground; and the very wild hollies, in hot aspects, were so much affected that they cast all their leaves.

By the 14th of January the snow was entirely gone; the turnips emerged not damaged at all, save in sunny places; the wheat looked delicately, and the garden plants were well preserved; for snow is the most kindly mantle that infant vegetation can be wrapped in; were it not for that friendly meteor no vegetable life could exist at all in northerly regions. Yet in Sweden the earth in April is not divested of snow for more than a fortnight before the face of the country is covered with flowers.

LETTER LXII

THERE were some circumstances attending the remarkable frost in January 1776 so singular and striking, that a short detail of them may not be unacceptable.

The most certain way to be exact will be to copy the passages from my journal, which were taken from time to time as things occurred. But it may be proper previously to remark that the first week in January was uncommonly wet, and drowned with vast rains from every quarter: from whence may be inferred, as there is great reason to believe is the case, that intense frosts seldom take place till the earth is perfectly glutted and chilled with water;* and hence dry autumns are seldom followed by rigorous winters.

January 7th. – Snow driving all the day, which was followed by frost, sleet, and some snow, till the 12th, when a prodigious mass overwhelmed all the works of men, drifting over the tops of the gates and filling the hollow lanes.

On the 14th the writer was obliged to be much abroad; and thinks he never before or since has encountered such rugged Siberian weather. Many of the narrow roads were now filled above the tops of the hedges; through which the snow was driven into most romantic and grotesque shapes, so striking to the imagination as not to be seen without wonder and pleasure. The poultry dared not to stir out of their roosting-places; for cocks and hens are so dazzled and confounded by the glare of snow that they would soon perish without assistance. The hares also lay sullenly in their seats, and would not move until compelled by hunger; being conscious, poor animals, that the drifts and heaps treacherously betray their footsteps, and prove fatal to numbers of them.

From the 14th the snow continued to increase, and began to stop the road waggons and coaches, which could no longer keep on their regular stages; and especially on the western roads, where the fall appears to have been deeper than in the south. The company at Bath, that wanted to attend the Queen's birth-day, were strangely incommoded: many carriages of persons, who got, in their way to town from Bath, as far as Marlborough, after strange embarrassments, here met with a *ne plus ultra*. The ladies fretted, and offered large rewards to labourers, if they would shovel them a track to London; but the relentless heaps of snow were too bulky to be removed; and

* The autumn preceding January 1768 was very wet, and particularly the month of September, during which there fell at Lyndon, in the county of Rutland, six inches and an half of rain. And the terrible long frost of 1739-40 set in after a rainy season, and when the springs were very high.

so the 18th passed over, leaving the company in very uncomfortable circumstances at the Castle and other inns.

On the 20th the sun shone out for the first time since the frost began; a circumstance that has been remarked before much in favour of vegetation. All this time the cold was not very intense, for the thermometer stood at 29, 28, 25, and thereabout; but on the 21st it descended to 20. The birds now began to be in a very pitiable and starving condition. Tamed by the season, skylarks settled in the streets of towns, because they saw the ground was bare; rooks frequented dunghills close to houses; and crows watched horses as they passed, and greedily devoured what dropped from them; hares now came into men's gardens, and, scraping away the snow, devoured such plants as they could find.

On the 22nd the author had occasion to go to London through a sort of Laplandian-scene, very wild and grotesque indeed. But the metropolis itself exhibited a still more singular appearance than the country; for, being bedded deep in snow, the pavement of the streets could not be touched by the wheels or the horses' feet, so that the carriages ran about without the least noise. Such an exception from din and clatter was strange, but not pleasant; it seemed to convey an uncomfortable idea of desolation:

> . . . ipsa silentia terrent.

On the 27th much snow fell all day, and in the evening the frost became very intense. At South Lambeth, for the four following nights, the thermometer fell to 11, 7, 6, 6; and at Selborne to 7, 6, 10; and on the 31st January, just before sunrise, with rime on the trees and on the tube of the glass, the quicksilver sunk exactly to zero, being 32 degrees below the freezing point; but by eleven in the morning, though in the shade, it sprung up to $16\frac{1}{2}$* – a most unusual degree of cold

* At Selborne the cold was greater than at any other place that the author could hear of with certainty: though some reported at the time

this for the south of England! During these four nights the cold was so penetrating that it occasioned ice in warm chambers and under beds; and in the day the wind was so keen that persons of robust constitutions could scarcely endure to face it. The Thames was at once so frozen over both above and below bridge that crowds ran about on the ice. The streets were now strangely incumbered with snow, which crumbled and trod dusty; and, turning grey, resembled bay-salt; what had fallen on the roofs was so perfectly dry that, from first to last, it lay twenty-six days on the houses in the city; a longer time than had been remembered by the oldest housekeepers living. According to all appearances we might now have expected the continuance of this rigorous weather for weeks to come, since every night increased in severity; but behold, without any apparent cause, on the 1st of February a thaw took place, and some rain followed before night; making good the observation above, that frosts often go off as it were at once, without any gradual declension of cold. On the 2nd of February the thaw persisted; and on the 3rd swarms of little insects were frisking and sporting in a court-yard at South Lambeth, as if they had felt no frost. Why the juices in the small bodies and smaller limbs of such minute beings are not frozen is a matter of curious inquiry.

Severe frosts seem to be partial, or to run in currents; for, at the same juncture, as the author was informed by accurate correspondents, at Lyndon in the county of Rutland, the thermometer stood at 19: at Blackburn, in Lancashire, at 19: and at Manchester at 21, 20, and 18. Thus does some unknown circumstance strangely overbalance latitude, and render the cold sometimes much greater in the southern than in the northern parts of this kingdom.

---

that at a village in Kent, the thermometer fell two degrees below zero, viz., 34 degrees below the freezing point.

The thermometer used at Selborne was graduated by Benjamin Martin.

The consequences of this severity were, that in Hampshire, at the melting of the snow, the wheat looked well, and the turnips came forth little injured. The laurels and laurustines were somewhat damaged, but only in hot aspects. No evergreens were quite destroyed; and not half the damage sustained that befell in January, 1768. Those laurels that were a little scorched on the south-sides were perfectly untouched on their north-sides. The care taken to shake the snow day by day from the branches seemed greatly to avail the author's evergreens. A neighbour's laurel-hedge, in a high situation, and facing to the north, was perfectly green and vigorous; and the Portugal laurels remained unhurt.

As to the birds, the thrushes and blackbirds were mostly destroyed; and the partridges, by the weather and poachers, were so thinned that few remained to breed the following year.

LETTER LXIII

As the frost in December, 1784, was very extraordinary, you, I trust, will not be displeased to hear the particulars; and especially when I promise to say no more about the severities of winter after I have finished this letter.

The first week in December was very wet, with the barometer very low. On the 7th, with the barometer at 28-five-tenths, came on a vast snow, which continued all that day and the next, and most part of the following night; so that by the morning of the 9th the works of men were quite overwhelmed, the lanes filled so as to be impassable, and the ground covered twelve or fifteen inches without any drifting. In the evening of the 9th the air began to be so very sharp that we thought it would be curious to attend to the motions of a thermometer:

we therefore hung out two; one made by Martin and one by Dollond, which soon began to show us what we were to expect; for, by ten o'clock, they fell to 21, and at eleven to 4, when we went to bed. On the 10th, in the morning, the quicksilver of Dollond's glass was down to half a degree below zero; and that of Martin's which was absurdly graduated only to four degrees above zero, sunk, quite into the brass guard of the ball; so that when the weather became most interesting this was useless. On the 10th, at eleven at night, though the air was perfectly still, Dollond's glass went down to one degree below zero! This strange severity of the weather made me very desirous to know what degree of cold there might be in such an exalted and near situation as Newton. We had therefore, on the morning of the 10th, written to Mr —, and entreated him to hang out his thermometer, made by Adams; and to pay some attention to it morning and evening; expecting wonderful phænomena, in so elevated a region, at two hundred feet or more above my house. But, behold! on the 10th, at eleven at night, it was down only to 17, and the next morning at 22, when mine was at 10. We were so disturbed at this unexpected reverse of comparative local cold, that we sent one of my glasses up, thinking that of Mr — must, somehow, be wrongly constructed. But, when the instruments came to be confronted, they went exactly together: so that, for one night at least, the cold at Newton was 18 degrees less than at Selborne; and, through the whole frost, 10 or 12 degrees; and indeed, when we came to observe consequences, we could readily credit this; for all my laurustines, bays, ilexes, arbutuses, cypresses, and even my Portugal laurels,* and (which occasions more regret) my fine sloping laurel hedge, were scorched up; while, at Newton, the same trees have not lost a leaf!

*Mr Miller, in his Gardener's Dictionary, says positively that the Portugal laurels remained untouched in the remarkable frost of 1739–40. So that either that accurate observer was much mistaken, or else the frost of December, 1784, was much more severe and destructive than that in the year above mentioned.

We had steady frost on to the 25th, when the thermometer in the morning was down to 10 with us, and at Newton only to 21. Strong frost continued till the 31st, when some tendency to thaw was observed, and, by January the 3rd, 1785, the thaw was confirmed, and some rain fell.

A circumstance that I must not omit, because it was new to us, is, that on Friday, December the 10th, being bright sunshine, the air was full of icy *spiculæ*, floating in all directions, like atoms in a sun-beam let into a dark room. We thought them at first particles of the rime falling from my tall hedges; but were soon convinced to the contrary, by making our observations in open places where no rime could reach us. Were they watery particles of the air frozen as they floated; or were they evaporations from the snow frozen as they mounted?

We were much obliged to the thermometers for the early information they gave us: and hurried our apples, pears, onions, potatoes, etc., into the cellar, and warm closets; while those who had not, or neglected such warnings, lost all their stores of roots and fruits, and had their very bread and cheese frozen.

I must not omit to tell you that, during those two Siberian days, my parlour-cat was so electric, that had a person stroked her, and been properly insulated, the shock might have been given to a whole circle of people.

I forgot to mention before, that, during the two severe days, two men, who were tracing hares in the snow, had their feet frozen; and two men, who were much better employed, had their fingers so affected by the frost, while they were thrashing in a barn, that a mortification followed, from which they did not recover for many weeks.

This frost killed all the furze and most of the ivy, and in many places stripped the hollies of all their leaves. It came at a very early time of the year, before old November ended; and yet it may be allowed from its effects to have exceeded any since 1739–40.

LETTER LXIV

As the effects of heat are seldom very remarkable in the northerly climate of England, where the summers are often so defective in warmth and sunshine as not to ripen the fruits of the earth so well as might be wished, I shall be more concise in my account of the severity of a summer season, and so make a little amends for the prolix account of the degrees of cold, and the inconveniences that we suffered from late rigorous winters.

The summers of 1781 and 1783 were unusually hot and dry; to them therefore I shall turn back in my journals, without recurring to any more distant period. In the former of these years my peach and nectarine-trees suffered so much from the heat that the rind on the bodies was scalded and came off; since which the trees have been in a decaying state. This may prove a hint to assiduous gardeners to fence and shelter their wall-trees with mats or boards, as they may easily do, because such annoyance is seldom of long continuance. During that summer also, I observed that my apples were coddled, as it were, on the trees; so that they had no quickness of flavour, and would not keep in the winter. This circumstance put me in mind of what I have heard travellers assert, that they never ate a good apple or apricot in the south of Europe, where the heats were so great as to render the juices vapid and insipid.

The great pests of a garden are wasps, which destroy all the finer fruits just as they are coming into perfection. In 1781 we had none; in 1783 there were myriads; which would have devoured all the produce of my garden, had not we set the boys to take the nests, and caught thousands with hazel twigs tipped with bird-lime: we have since employed the boys to take and destroy the large breeding wasps in the spring. Such expedients have a great effect on these marauders, and will keep them under. Though wasps do not abound but in hot summers, yet they do not prevail in every hot summer, as I have instanced in the two years above mentioned.

In the sultry season of 1783 honey-dews were so frequent as to deface and destroy the beauties of my garden. My honey-suckles, which were one week the most sweet and lovely objects that the eye could behold, became the next the most loathsome; being enveloped in a viscous substance, and loaded with black aphides, or smother-flies.[178] The occasion of this clammy appearance seems to be this, that in hot weather the effluvia of flowers in fields and meadows and gardens are drawn up in the day by a brisk evaporation, and then in the night fall down again with the dews, in which they are entangled; that the air is strongly scented, and therefore impregnated with the particles of flowers in summer weather, our senses will inform us; and that this clammy sweet substance is of the vegetable kind we may learn from bees, to whom it is very grateful: and we may be assured that it falls in the night, because it is always seen first in warm still mornings.

On chalky and sandy soils, and in the hot villages about London, the thermometer has been often observed to mount as high as 83 or 84; but with us, in this hilly and woody district, I have hardly ever seen it exceed 80; nor does it often arrive at that pitch. The reason, I conclude, is, that our dense clayey soil, so much shaded by trees, is not so easily heated through as those above-mentioned: and, besides, our mountains cause currents of air and breezes; and the vast effluvia from our woodlands temper and moderate our heats.

LETTER LXV

THE summer of the year 1783 was an amazing and portentous one, and full of horrible phænomena; for besides the alarming meteors and tremendous thunder-storms that affrighted and distressed the different counties of this kingdom, the peculiar haze, or smokey fog,[179] that prevailed for many weeks in this island, and in every part of Europe, and even beyond its limits, was a most extraordinary appearance, unlike anything known within the memory of man. By my journal I find that I had noticed this strange occurrence from June 23 to July 20 inclusive, during which period the wind varied to every quarter without making any alteration in the air. The sun, at noon, looked as blank as a clouded moon, and shed a rust-coloured ferruginous light on the ground, and floors of rooms; but was particularly lurid and blood-coloured at rising and setting. All the time the heat was so intense that butchers' meat could hardly be eaten on the day after it was killed; and the flies swarmed so in the lanes and hedges that they rendered the horses half frantic, and riding irksome. The country people began to look with a superstitious awe at the red, louring aspect of the sun; and indeed there was reason for the most enlightened person to be apprehensive; for, all the while, Calabria and part of the isle of Sicily, were torn and convulsed with earthquakes; and about that juncture a volcano sprung out of the sea on the coast of Norway. On this occasion Milton's noble simile of the sun, in his first book of *Paradise Lost*, frequently occurred to my mind; and it is indeed particularly applicable, because, towards the end, it alludes to a superstitious kind of dread, with which the minds of men are always impressed by such strange and unusual phænomena.

... As when the sun, new risen,
Looks through the horizontal, misty air,

Shorn of his beams; or from behind the moon,
In dim eclipse, disastrous twilight sheds
On half the nations, and with fear of change
Perplexes monarchs. . . .

LETTER LXVI

WE are very seldom annoyed with thunder-storms; and it is no
less remarkable than true, that those which arise in the south
have hardly been known to reach this village; for before they
get over us, they take a direction to the east or to the west, or
sometimes divide into two, and go in part to one of those
quarters, and in part to the other; as was truly the case in summer 1783, when though the country round was continually
harassed with tempests and often from the south, yet we escaped them all; as appears by my journal of that summer. The
only way that I can at all account for this fact – for such it is –
is that, on that quarter, between us and the sea, there are continual mountains, hill behind hill, such as Nore-hill, the Barnet,
Butser-hill, and Ports-down, which somehow divert the storms,
and give them a different direction. High promontories, and
elevated grounds, have always been observed to attract
clouds and disarm them of their mischievous contents, which
are discharged into the trees and summits as soon as they come
in contact with those turbulent meteors; while the humble
vales escape, because they are so far beneath them.

But, when I say I do not remember a thunder-storm from
the south, I do not mean that we never have suffered from
thunder-storms at all; for on June 5th, 1784, the thermometer
in the morning being at 64, and at noon at 70, the barometer at
29, six-tenths one-half, and the wind north, I observed a blue

mist, smelling strongly of sulphur, hanging along our sloping woods, and seeming to indicate that thunder was at hand. I was called in about two in the afternoon, and so missed seeing the gathering of the clouds in the north; which they who were abroad assured me had something uncommon in its appearance. At about a quarter after two the storm began in the parish of Hartley, moving slowly from north to south; and from thence it came over Norton-farm, and so to Grange-farm, both in this parish. It began with vast drops of rain, which were soon succeeded by round hail, and then by convex pieces of ice, which measured three inches in girth. Had it been as extensive as it was violent, and of any continuance (for it was very short), it must have ravaged all the neighbourhood. In the parish of Hartley it did some damage to one farm; but Norton, which lay in the centre of the storm, was greatly injured; as was Grange, which lay next to it. It did but just reach to the middle of the village, where the hail broke my north windows, and all my garden-lights and hand-glasses, and many of my neighbours' windows. The extent of the storm was about two miles in length and one in breadth. We were just sitting down to dinner; but were soon diverted from our repast by the clattering of tiles and the jingling of glass. There fell at the same time prodigious torrents of rain on the farms above-mentioned, which occasioned a flood as violent as it was sudden; doing great damage to the meadows and fallows, by deluging the one and washing away the soil of the other. The hollow lane towards Alton was so torn and disordered as not to be passable till mended, rocks being removed that weighed 200 weight. Those that saw the effect which the great hail had on ponds and pools say that the dashing of the water made an extraordinary appearance, the froth and spray standing up in the air three feet above the surface. The rushing and roaring of the hail, as it approached, was truly tremendous.

Though the clouds at South Lambeth, near London, were at that juncture thin and light, and no storm was in sight, nor within hearing, yet the air was strongly electric; for the bells of an electric machine at that place rang repeatedly, and fierce sparks were discharged.

When I first took the present work in hand I proposed to have added an *Annus Historico-naturalis*, or the Natural History of the Twelve Months of the Year; which would have comprised many incidents and occurrences that have not fallen in my way to be mentioned in my series of letters; but, as Mr Aikin of Warrington has lately published somewhat of this sort, and as the length of my correspondence has sufficiently put your patience to the test, I shall here take a respectful leave of you and natural history together; and am,

With all due deference and regard,

Your most obliged,

And most humble servant,

GIL. WHITE

Selborne,
June 25, 1787.

NOTES

1. (p. 7) 'Letter I'. The first nine letters to Thomas Pennant are not dated, and it is probable that White put them together just before his manuscript went to press. He needed some kind of introduction to describe the scene in which the book is set, and chose to disguise this as correspondence in the style of the remainder of the book.

2. (p. 9) 'Lime and turnips'. Turnips were introduced into the crop rotation system in the middle of the eighteenth century, as a source of winter forage and – as here – a way of clearing land of weeds. They revolutionized arable farming on poor soils.

3. (p. 10) 'or wych hazel'. Probably a wych elm, *Ulmus glabra*, native in northern Britain but planted elsewhere. This specimen, from the measurements White gives, could have been 250 years old.

4. (p. 11) 'Raven-tree'. Ravens ceased to nest in central southern England in the nineteenth century, though they are still to be found to the west, in Devon and Cornwall.

5. (p. 12) 'and by those who make collections *cock's comb*'. This fossil is *Ostraea ricordeana*, not Linnaeus' *Mytilus cristagalli*.

6. (p. 12) 'museum at Leicester-house'. Sir Ashton Lever's museum. This was one of the largest private collections ever assembled, and when it was finally dispersed in 1806, the auction lasted sixty-five days.

7. (p. 12) '*Cornua Ammonis*'. Common ammonites.

8. (p. 13) 'the freestone'. Greensand.

9. (p. 16) 'with their curious *filices*'. An early term for ferns.

10. (p. 16) 'after harvest some few landrails are seen'. Landrails (now known as corncrakes) and quails were apparently already beginning to decline in White's time as a result of the revolution in agricultural practices. Both are summer visitors, nesting in arable fields and meadows, and there is little doubt that new crops and cutting practices have affected their distribution. The corncrake, for instance, has virtually vanished from lowland Britain, where the hay is mown up to a month earlier than in the last century, and the birds robbed of cover during incubation.

11. (p. 20) 'without having one standing tree in the whole extent'. White meant 'Forest' in the ancient sense of a preserved hunting estate. These days much of Woolmer Forest is afforested in a more conventional sense.

NOTES

12. (p. 20) 'Dr Plot'. The curator of the Ashmolean Museum, Oxford, whose book was published in 1686.

13. (p. 20) 'or some such instruments'. In Ireland, and probably elsewhere, oak trunks were sometimes deliberately buried in bogs to mature and colour.

14. (p. 21) '*Hæmastatics*'. The second volume of *Vegetable Statics*; see note 145.

15. (p. 21) 'black game, or grouse'. The black grouse has continued to decline in Britain as a result of habitat destruction and over-shooting. It was extinct throughout Hampshire by 1920.

16. (p. 22) 'Waltham blacks'. These were described by the Draconian *Black Act*, as ill-designing and disorderly persons associated under the name of Blacks and 'entered into confederation to support and assist one another in stealing and destroying of deer, robbing of warrens and fish-ponds, cutting down plantations of trees, and other illegal practices, and have, in great numbers, armed with swords, fire-arms, and other offensive weapons, several of them with their faces blacked, or in disguised habits, unlawfully hunted in forests belonging to his Majesty, and in the parks of divers of his Majesty's subjects . . .'

17. (p. 25) 'feast of St Barnabas'. This was probably an ancient Midsummer ritual, as St Barnabas' day (11 June) was the longest day on the old calendar (though this is now ten or eleven days later; see note 130).

18. (p. 26) '*carex cespitosa*'. Probably *Carex paniculata*, otherwise known as tussock sedge. It is interesting that White uses the old word 'hassock' for tuft, as clumps of this sedge were once cut and used for church hassocks.

19. (p. 27) 'the term *purlieu*'. In this context, a tract of land on the border of a forest (or previously included in it), which was still subject to old forest laws.

20. (p. 30) 'lop and top'. An ancient common right to cut branches from timber trees for firewood.

21. (p. 31) 'worth attending to'. White returns time and time again to the subjects of migration and hibernation, and I have discussed his obsession with this subject in the Introduction. As to how his own and others' theories match up with our present state of knowledge, it has to be said that no evidence for the hibernation of swallows, or any other birds, has yet been discovered. But some individuals of various species have been found in torpid states in spells of very cold weather; and torpidity, involving a drastic slowing down of the whole metabolism, may be a kind of temporary hibernation, an adaptation to conditions when feeding is impossible.

White would also have been delighted to hear that some swallows are spending an increasing proportion of the winter in southern Britain, particularly in the warmth around coastal power stations.

22. (p. 32) '*motacilla trochilus*'. The wood warbler.

23. (p. 32) 'the fly-catcher'. The spotted flycatcher.

24. (pp. 32 & 95) 'They are delicate songsters'. It may be worth noting here that White failed to distinguish between the very similar songs of the blackcap and garden warbler. Indeed he makes no mention of the garden warbler at all. The bird was almost certainly quite common in Selborne (as it still is today), but it is secretive in its habits, and when White heard it singing from cover he presumably took it for a blackcap.

25. (p. 33) 'an excellent swimmer and diver'. This is almost certainly a water vole (still commonly known as the water rat), and White is correct in observing that its feet are not webbed. The other species of 'water-rat' which he saw in his stream may well have been the brown rat, which had arrived in this country from Russia about forty years before. It is a strong swimmer and was spreading rapidly across the country, ousting the native black rat as it went. It was certainly in Selborne ten years later, as White writes in his journal for 16 December 1777 that 'One black rat was killed at Shalden some months ago, and esteemed a great curiosity. The Norway rats destroy all the indigenous ones.' [1]

26. (p. 33) 'As to the *falco*'. This, as White was later told, was obviously a peregrine falcon. Hunting inland it may have been an escaped bird or an oddly marked juvenile. But it is still curious that White appeared to have difficulty in identifying it. Was he just charitably giving Pennant an opportunity to show his skill?

27. (p. 34) 'gross-beaks'. Hawfinches.

28. (p. 34) 'A cross-bill'. Crossbills have recently started to breed regularly in the New Forest (about thirty miles from Selborne), but their principal status in England is as winter visitors. About once every ten years they come in much greater numbers, and James Fisher has calculated that 1766 (the winter to which White is referring) could have been a suitable year for such an irruption.

29. (p. 35) 'two species of bats'. White almost certainly means the common pipistrelle, *Pipistrellus pipistrellus*, by his reference to '*vespertilio murinus*' (which is properly the similar but much scarcer parti-coloured bat), and the long-eared bat, *Pleucotus auritus*, by '*vespertilio auritus*'.

30. (p. 35) 'its wings round before the mouth'. Slow-motion film taken recently of hunting bats has shown that they also use their wings like this to scoop food out of the air.

31. (p. 36) 'the species is nondescript'. This is the harvest mouse, one

of White's original discoveries. His description has never been bettered, and is especially interesting because of its reference to the mice nesting in thistles. Until recently harvest mice were thought to nest almost exclusively in cornfields, and to be on the decline because of modern reaping techniques. It's now suspected that this was very much an 'over-flow' habitat, and that the mice are still quite common (though just as inconspicuous) in rank natural vegetation, reed-beds, scrub, etc.

32. (p. 37) 'German silk-tail'. The waxwing, which visits this country irregularly in winter (sometimes in irruptions) from northern Europe.

33. (p. 38) 'Flora'. An English translation of the *Calendarium Florae*, written by A. M. Berger, one of Linnaeus' students.

34. (p. 40) 'the bunting, *emberiza miliaria*'. The corn bunting.

35. (p. 40) 'Wagtails, both white and yellow'. Although White knew of the grey wagtail (it may even have bred in the streams round Selborne, as it does today) and distinguished it later (p. 111) from the yellow, he seems to be confusing the two species here. The yellow wagtail is a summer visitor, and leaves in October. The grey wagtail is present the whole year, and in Hampshire in winter is boosted by immigrants from the hill streams of north and west Britain.

The white wagtail is now known as the pied (*Motacilla alba*), and we reserve this English name for the paler race of *M. alba* which breeds in northern Europe and occurs in Britain as a passage migrant.

36. (p. 41) 'Hasselquist'. Another pupil of Linnaeus, sent by him to study the natural history of the Middle East.

37. (p. 42) 'the smallest quadrupeds in this island'. The smallest of all is in fact the pygmy shrew, which is often only half the weight of a harvest mouse, and two centimetres shorter.

38. (p. 42) 'two *spiracula*, or breathing-places'. Actually these are musk glands.

39. (p. 44) 'a *cane*'. The cane is really the female weasel, which is only four-fifths the length of the male. The polecat is now extinct in southern England.

40. (p. 44) 'the snow-flake'. The snow-bunting, a winter visitor to coastal areas and high ground.

41. (p. 45) 'never continuing one moment in the same place'. This description fits the siskin, a small and active finch which is now a regular breeder in Hampshire, and other regions of England where there are extensive conifer plantations. In White's time, though, it was probably just a winter visitor from the Continent. It is not strictly a soft-bill (in White's sense of an insect-eater), but a seed-eater.

42. (p. 46) 'three species of the willow-wrens'. White was the first observer to clearly differentiate our three common leaf warblers. The

songs he describes are still the best way of distinguishing the birds in the field (and is probably the way the birds themselves tell each other apart!).

Since these species are mentioned again several times in the book, here is a key to White's song descriptions and names: Smallest willow-wren, with 'a harsh loud chirp' – our chiffchaff, *Phylloscopus collybita*. Middle willow-wren, with 'a joyous, easy, laughing note' – the willow warbler, *Phylloscopus trochilus*. Largest willow-wren, which makes 'a sibilous shivering noise' – the wood warbler, *Phylloscopus sibilatrix*.

43. (p. 46) 'the grasshopper-lark'. The grasshopper warbler.

44. (p. 48) 'Some suspect it is made by their wings'. A hundred years later it was proved that the 'drumming' of the snipe is produced by rapid vibrations in the bird's outer tail feathers, which are thrust out at right angles to the body during the fast downward plunges in its display flight.

45. (p. 48) 'the golden-crowned wren'. The goldcrest.

46. (p. 49) 'Swammerdam'. J. J. Swammerdam (1637–80), a specialist on amphibians.

47. (p. 49) 'toads being observed in the same situation'. It is odd that White never saw this, as male toads frequently mount the females during their laborious journeys to the spawning grounds. Like frogs, toads are oviparous, the female laying eggs on water plants, and her partner fertilizing them *in situ*.

The so-called venom is in fact a defensive slime which is highly distasteful to predatory animals. Interestingly – given White's later account of 'toad-cures' (page 53) – one of its active principles, called *bufotenin*, has been used experimentally in the treatment of schizophrenia.

48. (p. 50) 'the tail drops off as useless'. It is absorbed more than discarded.

49. (p. 50) '*rana arborea* is an English reptile'. The European tree frog is certainly not an indigenous English amphibian (it is not strictly a reptile), though captive specimens have occasionally been successfully introduced into the wild. But, as White suspected, Christopher Merret (1619–95) was not to be trusted as a reporter. His *Pinax Rerum Naturalium* is an inaccurate and jumbled list of British animals, vegetables and minerals.

50. (p. 50) '(the coralline Ellis)'. So-called because he demonstrated the animal nature of coral.

51. (p. 50) 'the water-eft, or newt'. White refers here to the crested newt, *Triturus cristatus*, which is not the larva of any kind of 'land-eft' or lizard. There was a good deal of confusion in his time between lizards and newts. Lizards are true reptiles, and at all stages of their life are independent of water. Newts, like frogs, are amphibians, pass

through a water-borne tadpole stage, and spend much of their adult life in the water. The confusion was partly a result of the male newt losing its crest for the period it lives out of water – thus looking as well as behaving more like a lizard.

52. (p. 51) 'Whereas snakes lay a chain of eggs'. That is, grass-snakes (which White refers to as 'common snakes' later in the paragraph), not vipers.

53. (p. 52) 'or rather varieties, of our *lacerti*'. There are only two species of lizard ('*lacerti*') native to the British Isles, the common and the sand lizard, though males of the latter species are often coloured green. The true green lizard, *Lacerta viridis*, has occasionally been introduced from France and the Channel Islands. The total number of indigenous British reptiles is thus five: two lizards and three snakes.

54. (p. 52) 'no such fish as the *gasterosteus pungitius*'. This is the rare ten-spined stickleback. G. *aculeatus*, the three-spined, is much commoner.

55. (p. 55) 'red-backed butcher bird'. It is curious that some writers classed the red-backed shrike with birds 'only to be seen in the northern counties', and that White did not discover it till 1768. Although it is now declining fast as a British breeding species, it has always been a bird of the warmer southern counties; and the rough heaths of Woolmer Forest should have been ideal for it.

56. (p. 56) 'a large black warty lizard'. The great crested newt, which is capable of scaling moist walls.

57. (p. 57) 'Mr.——'. Almost certainly John Hunter, who was rising to fame as an anatomist and had just been elected an F.R.S.

58. (p. 57) 'the *Naturalist's Journal*'. The specially printed diaries for recording natural history observations, which are mentioned in the Introduction.

59. (p. 58) 'never straggle to you'. Pennant lived in Flintshire, way outside the stone curlew's breeding range.

60. (p. 59) 'meanly furnished with churches'. A remark which says more about White's limited travels than the state of ecclesiastical building.

61. (p. 60) 'do certainly not forsake them against winter'. Pennant was mistaken. The ring ouzel migrates to Africa for the winter, passing through the southern counties in September and October, and back again in early spring. White suspected this from his own observations and reasoning, and was never happy with Pennant's conclusion. He worries at the subject again and again, and the cautious, schoolmasterly tone of his injunction to Pennant later in this paragraph becomes plainly sceptical by Letter XXXI (page 79).

62. (p. 60) '(the goat-sucker)'. The nightjar. The chattering in flight is produced by fast wing-clapping.

63. (p. 64) 'The *scarabaeus fullo*'. A cockchafer beetle not found in Britain.

64. (p. 64) 'Mr Banks'. Sir Joseph Banks, F.R.S. (1743–1820), the distinguished naturalist who accompanied Captain Cook on his expeditions. Both White and Pennant met him.

65. (p. 65) 'a new *salicaria*'. This is the sedge warbler, and the description which follows is the first accurate one to appear in print.

66. (p. 65) 'a reed-sparrow'. Probably the reed bunting, not a warbler.

67. (p. 69) '*in scapulis alba Raii*'. The woodchat-shrike, a rare vagrant to Britain from southern Europe.

68. (p. 70) 'never choose to breed in England'. This may yet prove a prophetic reflection, as a pair of fieldfares bred in the Orkneys in 1967, and there are signs that it is becoming a regular British breeding species.

69. (p. 71) 'The eagle-owl'. Only a very rare vagrant to Britain.

70. (p. 72) 'supported itself for the winter'. This second-hand story of a '*hybernaculum*' with a store of food suggests a brown rat. Water voles do not lay up food, nor do they stray far from water. It is marvellous to see White's intuitive intelligence sensing something wrong here, and working with the evidence.

71. (pp. 72 & 86) 'The great large bat'. This is certainly the noctule bat, *Nyctalus noctula*. It was first described ten years before by the French naturalist Daubenton, but White must be given credit for its discovery in Britain. The 'little bat' in the footnote is the pipistrelle.

72. (p. 72) 'curlews'. That is, stone curlews. White never mentions the common curlew.

73. (p. 76) 'the same creature'. Moose and elk are the same species.

74. (p. 80) '... *fugiens ab hoste*'. Translated: 'It carries its young in its beak when escaping from an enemy.' White was correct to mistrust this statement as it stands. What has been proved, though, is that the woodcock occasionally carries one chick at a time between its thighs.

75. (p. 80) '*hirundo rupestris*'. The mediterranean crag martin, which has never been recorded in Britain.

76. (p. 82) '*hirundo melba*, the great Gibraltar swift'. This is the Alpine swift, now classified as *Apus melba*, which is common in southern Europe, and an occasional straggler to Britain.

77. (p. 83) 'my relation'. His brother, John.

78. (p. 84) '(which we call an harvest-bug)'. The red larva of the mite *Trombicula autumnalis*, which attacks almost every kind of animal in late summer. Curiously, the adult stage has never been seen.

79. (p. 84) 'small long shining fly'. The cheese-fly, *Piophila casei*.

80. (p. 84) 'turnip-fly and black dolphin'. Now rather less graphically known as flea-beetles, or leaf-beetles (*Chrysomelidae* family).

81. (p. 85) 'legs and flanks of grass-horses'. The horse bot-fly, *Gastrophilus equi*, which lays its eggs on hairs on the horse's forequarters. When they hatch out, the irritation causes the horse to lick the hairs and swallow the larvae. These larvae feed on the inner lining of the stomach until they pass out with the dung a year later. They then bury themselves in the ground and pupate. 'Old Moufet' was Dr Thomas Moufet, or Muffet, the Elizabethan author of *A Theatre of Insects*.

82. (p. 85) '*musca chamaeleon*'. The larva of the fly *Stratiomys chamaeleon*.

83. (p. 87) 'a peculiar structure'. This is the tragus, or false inner ear. Its precise function isn't properly understood, but it probably plays a part in the echo-sounding system with which bats find their way about.

84. (p. 88) 'a serrated claw'. This middle claw is used solely for preening, and not, as White surmised, for delivering food to the mouth.

85. (p. 90) 'a great ash-coloured butcher bird'. The great grey shrike, now a regular winter visitor to Britain.

86. (p. 90) 'Cornish choughs'. Choughs have vanished not only from Sussex, but, since 1952, from Cornwall. They are now confined chiefly to the west coasts of Wales and Ireland.

87. (p. 90) 'wild-pigeon, or stock-dove, is a bird of passage'. Stock-doves were almost certainly regular breeders in the south of England, then as now; and though they don't indulge in mass migrations, woodpigeons do. To add to the confusion, White called stock-doves woodpigeons, and wood pigeons ring-doves. See page 102.

88. (p. 91) 'Titlarks'. White did not distinguish between the very similar tree and meadow pipits. This is a clear description of the songflight of the tree pipit.

89. (p. 91) 'Whin-chats'. Whinchats are summer visitors and are absent from the British Isles between October and April. White may have misidentified juvenile and female stonechats (residents) in their winter plumage.

90. (p. 93) 'the generation of eels is very dark and mysterious'. Eels breed only in the Gulf of Mexico, and the threads referred to by White are parasitic worms.

91. (p. 94) 'young cock red-breasts of that year'. Though robins are one of the few species where the females also hold territory and sing during the winter.

92. (p. 94) 'the marsh titmouse'. This is a confusing passage. Two 'quaint notes, like the whetting of a saw' is a fairly broad description,

and most of the tit family include calls like this amongst their extensive repertoires. It is the commonest call of the great tit (though this bird has more than thirty distinct calls, including White's 'three cheerful joyous notes'), and also occurs in the willow tit. This bird is so similar in appearance to the marsh tit that White did not realize they were two distinct species, and it was not until the end of the nineteenth century that they were properly separated.

93. (p. 95) 'grey crows'. Hooded crows, winter visitors to southern Britain.

94. (p. 95) '*scarabaeus solstitialis*, or fern-chafer'. The summer chafer, *Rhizotragus solstitialis*. Nightjars also feed on many other species of dusk-flying beetles and moths.

95. (p. 96) 'This bird does not make the least pretension to song'. In fact the spotted flycatcher has a faint, squeaky song.

96. (p. 96) '252 species'. There are now more than 500 species on the British list, if all the rare vagrants are included. The increase is probably due as much to the proliferation of sharp-eyed latter-day Whites as to real changes in bird distribution.

97. (p. 97) 'imbecility'. That is, physical weakness.

98. (p. 97) 'their *aurelia* state'. The pupa, or chrysalis stage.

99. (p. 99) 'some plants little to be expected'. Another prophetic remark. A group of plants found growing together principally in the Iberian peninsula, and collectively known as the Lusitanian flora, were discovered later in south-western Ireland.

100. (p. 100) 'honey-buzzards'. In White's time a regular breeding bird in the south of Britain. Another pair raised young in the Hanger in 1787. But by the end of the nineteenth century the bird was almost extinct, and it must be said that egg collecting of the kind described in this letter played a role in its decline. Happily it now seems to be increasing again, and a handful of pairs breed annually, chiefly in White's own county.

101. (p. 110) 'less reed-sparrow'. The sedge warbler.

102. (p. 111) 'Royston-crow'. The hooded crow.

103. (p. 112) 'Wild-duck'. It's not possible to identify White's 'Wild-goose' exactly, but the 'Wild-duck' was almost certainly the mallard.

104. (p. 115) 'bunting'. The corn bunting.

105. (p. 118) 'no one supposes they sing in the night'. Robins do in fact sing quite commonly at night.

106. (p. 119) 'on the same food with the hard-billed'. The young of most species of bird, whether hard- or soft-billed, are fed on insects.

107. (p. 121) 'thick-billed reed-sparrow'. The reed bunting.

108. (p. 121) 'the vast disproportioned size of supposititious egg'. Cuckoos' eggs are only slightly larger than their hosts', and are almost indistinguishable in colour.

109. (p. 122) 'drawing up an account of the animals in this neighbourhood'. The first reference to the idea of the *Natural History*.

110. (p. 126) 'birds of prey'. Linnaeus, and many other contemporary writers, classed cuckoos as birds of prey on no other grounds than their shape and barred breast feathers.

111. (p. 126) 'the water-ousel'. The dipper, which breeds in rocky mountain streams and visits river estuaries occasionally in winter.

112. (p. 127) 'bustards on the wide downs near Brighthelmstone'. Great bustards were soon to become extinct in Sussex, and had disappeared as a British breeding species by 1833. Brighthelmstone is Brighton.

113. (p. 127) 'a land-tortoise'. This is the famous Timothy, an Iberian tortoise which eventually found its way into White's possession, and is mentioned frequently by him in later letters and journals. His shell is preserved in the British Museum.

114. (p. 128) '*aberdavines*'. Siskins.

115. (p. 129) 'a woodcock's nest'. Woodcocks now breed more commonly in Hampshire.

116. (p. 129) 'Redwings'. Now breeding regularly in Scotland.

117. (p. 130) '*maximis in arboribus nidificat*'. Translated, these passages read: 'It nests in the highest trees'; and of the redwing, 'it nests in medium-sized bushes or hedges; eggs six, blue-green, varied with black spots.' Of the woodcock, Scopoli says that 'it nests in swampy Alpine woods; it lays 3–5 eggs.' It does not appear from Kramer that woodcocks breed at all in Austria; but he says, 'the bird is a summer inhabitant of northern districts, where it breeds in numbers. On the approach of winter it moves to southern lands; hence it passes through Austria plentifully about the October full moon. It returns again; preferably about the March full moon, mated, passing through Austria to northern districts.'

118. (p. 131) 'bee-birds ... golden thrushes'. Bee-eaters ... golden orioles.

119. (p. 133) '*alauda pratensis*'. The meadow pipit – White's titlark.

120. (p. 134) 'My musical friend'. His brother, the Rev. Henry White, rector of Fyfield, Hants.

121. (p. 135) '*ob eandem causam*'. Translated: 'Birds of the Rail family fly away as if by common consent; not a single one can be found remaining with us; in summer they cannot survive in southern districts

because of the disappearance of worms, and the parched earth; and so it is in cold regions, for the same reason.'

122. (p. 141) 'the blue thrush'. The rock thrush, which breeds around the Mediterranean and is a very rare vagrant to Britain.

123. (p. 142) 'white-bellied field-mouse'. The long-tailed field mouse.

124. (p. 146) *'hippoboscae hirundinis'*. *Stenopteryx hirundinis*, an extraordinary fly which lays, not eggs, but young at such an advanced stage that they immediately pupate.

125. (p. 146) 'Réaumur'. A distinguished French scientist, of obviously remarkable dedication, who lived from 1683–1757.

126. (p. 149) 'the dams'. In fact both sexes take equal parts in nest-building, incubation and the feeding of young.

127. (p. 151) 'your respectable society'. The letter referred to was read before the Royal Society on 10 February 1774.

128. (p. 153) 'The black-faced poll-sheep'. The Southdown sheep, ancestor of most of our modern black-faced sheep.

129. (p. 155) 'whin-chats, some rooks, and several kites and buzzards'. These were probably stonechats, as whinchats, being summer visitors, would have departed by this date. Red kites were soon to become extinct in England. They are now restricted to a population of some twenty-five pairs in central Wales.

130. (p. 156) 'early in February'. White's boyhood could be said to have stretched until 1740, and this reminiscence therefore refers to a period before the calendar revision of 1752, when the whole system was shifted eleven days forward. This early February date thus becomes a late February one – still remarkable for a swallow.

131. (p. 156) 'this *hirundo* breeds in chimneys'. It is worth remembering that cottage chimneys in White's day were very much more roomy than today, to cope with the smoke from bulky wood fires. These days swallows build chiefly on rafters in barns and outhouses.

132. (p. 162) *'Hirundo esculenta'*. This is the 'edible' swift, *Collocalia esculenta*, whose saliva-built nests are used to make 'bird's nest soup'. It is a true swift, which White incorrectly, though excusably, classed with the hirundines, i.e. martins and swallows.

133. (p. 164) 'bed fleas'. This is a species specific to the sand-martin, *Ceratophyllus styx*, and not the human flea, *Pulex irritans*.

134. (p. 165) 'rather mute'. In fact they have a slight, dry, chattering song.

135. (p. 166) 'a nice observer'. Gilbert's brother, John, again.

136. (p. 168) 'but two eggs at a time'. Actually the most usual number is three, though two and four are not uncommon. These take eighteen to nineteen days to hatch, and the young remain in the nest for about

six weeks. These lengthy periods underline the fact that swifts are in a different order (*Apodiformes*) from the martins and swallows (*Passeriformes*), which take only half this time to complete their breeding cycle. Unlike these true hirundines, incubation for swifts is largely the responsibility of the female, though the male will sometimes join her in the nest at night. More often he spends the entire night asleep on the wing.

137. (p. 170) 'retire to rest for a season'. It's conspicuous the way that White's normally sharp reasoning starts to falter whenever he tries to square up his observations with his deepening belief in bird hibernation. Swifts normally moult twice, once in the autumn and again in spring, before migration. They therefore arrive with us in brand-new dark plumage.

138. (p. 173) 'those of Tobit'. The quotation reads: 'The same night also I returned from the burial and slept by the wall of my courtyard, being polluted, and my face was uncovered. And I knew not that there were sparrows in the wall, and mine eyes being open, the sparrows muted warm dung into mine eyes, and I went to the physicians, but they helped me not.'

139. (p. 174) '*cuculus cuculat*'. That is, 'hawks make peace with the birds, as long as the cuckoo calls', a country saying probably originating in the frequent misidentification of insectivorous cuckoos as birds of prey.

140. (p. 176) 'a gentleman'. The gentleman about whom White speaks with such filial respect was, of course, his father, John White.

141. (p. 177) 'Dr Lister'. Dr Martin Lister, F.R.S., physician to Queen Anne.

142. (p. 179) 'Mezeray'. François Eudes de Mezeray (1610–83), a French historian, largely responsible for perpetuating the myth that gypsies originated in Egypt. Today it is believed that they migrated from India in the twelfth century.

143. (p. 183) 'metheglin'. A mead-like drink made from spiced, fermented honey.

144. (p. 187) 'alembics'. Devices for distillation.

145. (p. 189) 'Dr Hales'. The Rev. Stephen Hales (1677–1761), whose important and imaginative book *Vegetable Statics* was one of the first works in physiology. He was one of White's very few intellectual neighbours (he was rector of Faringdon, two miles from Selborne, for many years), and encouraged him to keep a naturalist's calendar.

146. (p. 189) 'a large quantity of moisture nightly by condensation'. Earth, in fact, cools off more quickly than water during the night, and soon drops to a temperature lower than that of the surface of the water. So most 'falling dew' condenses on the earth around a pond, not in the

NOTES

water itself. It's now believed that these curious 'dew-ponds' are filled almost exclusively by rain.

147. (p. 193) 'Mr Lisle'. Edward Lisle, of Crux-Easton, Hampshire, whose *Observations on Husbandry* was published posthumously in 1757. His observation on impotence following the removal of a boar's tusks has recently been confirmed. It seems as if horns and tusks – which have no apparent physiological function – are psychologically important to male animals' virility.

148. (p. 195) 'the *murium leo*'. That is, the lion of mice.

149. (p. 197) 'A good monography of worms'. Such a work was not to be produced until Darwin wrote *The Formation of Vegetable Mould through the Action of Worms*, in 1881. For some reason, Darwin does not mention White's observations, though he was well acquainted with, and an admirer of, the *Natural History*.

150. (p. 201) 'hinds'. Not a disparaging term for wives, but a name sometimes given to farm servants.

151. (p. 205) 'any organs of hearing at all'. On the contrary, insects have excellent organs of hearing, though these are often located in unexpected places such as the leg.

152. (p. 208) 'the young seek for new abodes'. Though the young of many species of migrant bird return to breed in the sites where they themselves were reared – that is, those that survive the winter.

153. (p. 210) 'champaign fields'. Simply, open meadowland.

154. (p. 211) 'narrow-leaved, or wild lathyrus'. Narrow-leaved everlasting pea, *Lathyrus sylvestris*.

155. (p. 212) '*Ophrys nidus avis*, birds' nest ophrys'. Birdsnest orchid, *Neottia nidus-avis*, not the same plant as the yellow birdsnest (White's yellow mototropa).

156. (p. 212) '*Serapias latifolia*, helleborine'. Probably the broad-leaved helleborine. *Epipactis helleborine*, the commonest member of the genus.

157. (p. 212) 'of all the propensities of plants'. The whole of the following paragraph is missing from the first edition.

158. (p. 212) 'only one species'. In fact there are four distinct purple-flowered species confused here. *Crocus sativus*, the true saffron crocus, is an autumn flowerer introduced from Asia by the Romans. It does not occur in the wild in Britain. Similar are the native meadow saffron, *Colchicum autumnale*, and the autumn crocus, *Crocus nudiflorus*, which, like the vernal or spring crocus, is a naturalized introduction from southern Europe. They all differ slightly in their internal structure.

159. (p. 213) 'et in terrâ, et in aëre'. Translated: 'All living creatures have a distinctive and individual style of movement, each according to

its kind; that of birds alone varies both on the ground and in the air.' Part of what modern bird-watchers call a species' 'jizz'.

160. (p. 214) 'scratching themselves with one foot'. Although ravens will occasionally scratch themselves in flight, this isn't the cause of their aerial rolls and dives, which are part of the male's display flight.

161. (p. 215) *'tringae'*. Waders of the sandpiper group.

162. (p. 219) 'Letter XLIV'. This and the succeeding undated letters were specially written to round off the book (cf. the first nine letters to Pennant), and may never have been sent to Barrington.

163. (p. 221) *'Baker's Chronicle'. A Chronicle of the Kings of England from the Time of the Romans' Government unto the Death of King James*, written by Sir Richard Baker (1568–1645) whilst he was in the Fleet Prison. In it he refers to an earthquake which occurred in eastern Herefordshire.

164. (p. 221) 'John Philips, in his *Cyder*'. A late-seventeenth-century bucolic poet. The Marcley Hill in White's extract from *Cyder* is the site of the earthquake in Baker's *Chronicle*.

165. (p. 230) 'chew the cud like many quadrupeds!'. White was right to be astonished. Mole crickets do not chew the cud.

166. (p. 231) *'charadrius himantopus'*. The blackwinged stilt, *Himantopus himantopus*. It is normally a very rare vagrant from southern Europe, but in 1945 there was an unprecedented influx, and three pairs nested on a sewage farm near Nottingham.

167. (p. 231) 'Brisson'. Mathurin Jacques Brisson (1723–1806), a tutor at the French court, and author of a large catalogue of birds.

168. (p. 236) *'coccus vitis viniferae'*. The beetle *Pulvinaria vitis*.

169. (p. 237) 'Mr Lightfoot'. John Lightfoot (1735–88), who wrote the celebrated *Flora Scotica* as a result of an expedition to Scotland with Pennant in 1772.

170. (p. 240) 'stews'. That is, stew-ponds, waters kept especially stocked with fish for eating.

171. (p. 244) 'a wimble'. A gimlet.

172. (p. 244) 'facultatem commovere'. Translated: 'Moreover he preferred the music of birds to human voices or musical instruments; not that the latter failed to please him; but human music disturbed his mind, claiming his attention and interrupting sleep with its rise and fall, and changes of pitch and time; but the warbling of birds left no such disturbance; for that which we cannot imitate cannot excite our minds either.'

173. (p. 245) 'the pettichaps'. The lesser whitethroat, *Sylvia curruca*, a summer visitor to Britain first described by Linnaeus in 1758.

174. (p. 246) 'In winter 1767'. That is, the 1766–7 winter.

175. (p. 250) 'Dr Huxham's *Observations de Aëre*'. Dr John Huxham (1692–1768), a meteorologist and a student of fever incidence.

176. (p. 251) 'My friend who lives just beyond the top of the down'. Another one of Gilbert's clerical relatives, his nephew the Rev. Edmund White.

177. (p. 254) 'Dr Fothergill'. Dr John Fothergill (1712–80), a Quaker physician who kept a large private botanic garden at Upton, Essex.

178. (p. 264) 'loaded with black aphides, or smother-flies'. Honey-dew is, in fact, a secretion produced by these aphids, and not the distilled essence of flowers.

179. (p. 265) 'the peculiar haze, or smokey fog'. Due to the eruption of the volcano Skaptár-jökull, in Iceland – one of the most cataclysmic in historic times.

READ MORE IN PENGUIN

In every corner of the world, on every subject under the sun, Penguin represents quality and variety – the very best in publishing today.

For complete information about books available from Penguin – including Puffins, Penguin Classics and Arkana – and how to order them, write to us at the appropriate address below. Please note that for copyright reasons the selection of books varies from country to country.

In the United Kingdom: Please write to *Dept. EP, Penguin Books Ltd, Bath Road, Harmondsworth, West Drayton, Middlesex UB7 0DA*

In the United States: Please write to *Consumer Sales, Penguin Putnam Inc., P.O. Box 999, Dept. 17109, Bergenfield, New Jersey 07621-0120.* VISA and MasterCard holders call 1-800-253-6476 to order Penguin titles

In Canada: Please write to *Penguin Books Canada Ltd, 10 Alcorn Avenue, Suite 300, Toronto, Ontario M4V 3B2*

In Australia: Please write to *Penguin Books Australia Ltd, P.O. Box 257, Ringwood, Victoria 3134*

In New Zealand: Please write to *Penguin Books (NZ) Ltd, Private Bag 102902, North Shore Mail Centre, Auckland 10*

In India: Please write to *Penguin Books India Pvt Ltd, 210 Chiranjiv Tower, 43 Nehru Place, New Delhi 110 019*

In the Netherlands: Please write to *Penguin Books Netherlands bv, Postbus 3507, NL-1001 AH Amsterdam*

In Germany: Please write to *Penguin Books Deutschland GmbH, Metzlerstrasse 26, 60594 Frankfurt am Main*

In Spain: Please write to *Penguin Books S. A., Bravo Murillo 19, 1° B, 28015 Madrid*

In Italy: Please write to *Penguin Italia s.r.l., Via Benedetto Croce 2, 20094 Corsico, Milano*

In France: Please write to *Penguin France, Le Carré Wilson, 62 rue Benjamin Baillaud, 31500 Toulouse*

In Japan: Please write to *Penguin Books Japan Ltd, Kaneko Building, 2-3-25 Koraku, Bunkyo-Ku, Tokyo 112*

In South Africa: Please write to *Penguin Books South Africa (Pty) Ltd, Private Bag X14, Parkview, 2122 Johannesburg*

READ MORE IN PENGUIN

A CHOICE OF CLASSICS

Matthew Arnold	**Selected Prose**
Jane Austen	**Emma**
	Lady Susan/The Watsons/Sanditon
	Mansfield Park
	Northanger Abbey
	Persuasion
	Pride and Prejudice
	Sense and Sensibility
William Barnes	**Selected Poems**
Anne Brontë	**Agnes Grey**
	The Tenant of Wildfell Hall
Charlotte Brontë	**Jane Eyre**
	Shirley
	Villette
Emily Brontë	**Wuthering Heights**
Samuel Butler	**Erewhon**
	The Way of All Flesh
Lord Byron	**Selected Poems**
Thomas Carlyle	**Selected Writings**
Arthur Hugh Clough	**Selected Poems**
Wilkie Collins	**Armadale**
	The Moonstone
	No Name
	The Woman in White
Charles Darwin	**The Origin of Species**
	Voyage of the *Beagle*
Benjamin Disraeli	**Sybil**
George Eliot	**Adam Bede**
	Daniel Deronda
	Felix Holt
	Middlemarch
	The Mill on the Floss
	Romola
	Scenes of Clerical Life
	Silas Marner

READ MORE IN PENGUIN

A CHOICE OF CLASSICS

READ MORE IN PENGUIN